What's Wrong With Our Technological Society— And How to Fix It

What's Wrong With Our Technological Society— And How to Fix It

Simon Ramo, Ph.D.

Chairman of the Board, the TRW-Fujitsu Company
Director, TRW Inc.
Visiting Professor, California Institute of Technology

McGraw-Hill Book Company

New York   St. Louis   San Francisco   Auckland
Bogotá   Hamburg   Johannesburg   London   Madrid
Mexico   Montreal   New Delhi   Panama   Paris
São Paulo   Singapore   Sydney   Tokyo   Toronto

Library of Congress Cataloging in Publication Data
Ramo, Simon.
    What's wrong with our technological society—and how to fix it.
    Includes index.
    1. Technological innovations—Social aspects—United States. I. Title.
T173.8.R35 1983   303.4'83   83-696
ISBN 0-07-051169-1

1234567890   DOC/DOC   89876543

ISBN 0-07-051169-1

The editors for this book were William A. Sabin and Olive H. Collen, the
designer was Dennis Sharkey, and the production supervisor was Teresa F.
Leaden. It was set in Garamond by Santype-Byrd.

Printed and bound by R. R. Donnelley & Sons Company.

Contents

Chapter One A Century Yet to Be Named

What beneficial advances or detrimental occurrences might future historians conclude shaped civilization most markedly in the twentieth century? The candidates worthy of consideration are great in number and diversity, from the coming of mass production to the advent of world wars, from penicillin to pollution, from communism to computers, from women's suffrage to outer space exploration. Even a violent restructuring of society by all-out nuclear war is not to be ruled out, since the awesome capability exists to make it happen and no comparably powerful wave of social maturation seems certain of arriving before the year 2000 to preclude it. Were this cataclysm to happen, civilization would not be wiped out. Billions of people would be left alive to continue on. But the effects would be noticeable for all the future. In their bodies and minds the survivors and their offspring would manifest evidence of what had transpired, and that event would dominate what would be said about this century ever after.

A full-scale nuclear war would not merely decimate the populations of the areas struck directly. Major portions of the earth would become uninhabitable for decades. Communication, transportation, production, energy, and hygenic resources would be reduced to small fractions of prewar strength, and uncontaminated food and water would be virtually unavailable in large parts of the planet precisely where populations of the technologically advanced nations were most dense before the bombs detonated. The remaining workable infrastructure there would not support the surviving society with its crippled, blind, and physically and mentally ill. Organized social frameworks would collapse as the temporarily still-active people competed for sustenance. After the catastrophe, a century or more might pass before a worldwide society would become operable again. Those born later would have a compelling, and not merely a scholarly, objective in seeking to understand us. They would hope through such comprehension to ensure that what had occurred would never be repeated.

Century of Mismatch

But still, what would they say about us? How would they describe the crux of our civilization's failure? Certainly they would conclude that our century suffered from a fatal imbalance between accelerating technological advance and lagging social progress. It was in the twentieth century, their analysis surely would point out, that we discovered the secrets of the inner workings of the atom and how to manipulate its constituents to produce energy releases powerful enough to wipe humankind and its manufactures from the face of the earth. They would pronounce it to have been the source of the disaster that homo sapiens attained these scientific and technological break-throughs before its separate tribes learned how to live peacefully on the same planet. Citing the critical significance of this imbalance, they might label ours the "century of mismatch," a period in time overwhelmed by the consequences of the race's failure to match technological strides with social advance.

If, despite the fact that a nuclear holocaust comes to characterize the twentieth century, a future society does indeed study this period, that suggests they would have recovered and found the time. But that does not mean they necessarily would have solved the mismatch problem. In their own way, they might emulate us and engage again in developing the technological means for their annihilation while once more failing to progress enough socially. Perhaps the more philosophical of their observers, puzzling over this continuing dyssymmetry between technological and social evolutions, might ask themselves if the species suffers from a fundamental defect—in essence, a curse. They might reason that humans on earth really are not representative of intelligent life, it being conceivably a universal law of the race that when it learns how to destroy itself, it does so soon after.

Fortunately, more optimistic possibilities can be suggested, even assuming the nuclear catastrophe happens. The deeply penalizing experience of our century might be exactly what the future population would need to rise to the challenge of making adequate social headway. A dark period of struggle to survive might come first, but then a reconstructed civilization based on superior human behavior might emerge. That civilization might figure out how to use science and technology solely for the benefit of humankind. If so, a golden age would follow.

Even if we don't annihilate ourselves before the end of this century, we are unlikely to celebrate at that time the total elimination of social instabilities, wars, economic disasters, poverty, starvation, and the present pattern in which much of our human and physical resources are devoted to arms. The twentieth century is probably not going to be described later as that era in history when the human race surmounted its problems. It still may deserve to be called the century of mismatch.

The First Technological Civilization

Our century might also be labeled the "century of technology," although obviously the twentieth will not stand out as the only century affected by technological advance. Technology has been with us since before the invention of the wheel, and the future society certainly will find itself under the spell of far more scientific discovery and technological development than we have so far known. But ours may go down as the century of technology because it will be seen that in the 1900s society did more than incorporate its share of technological advances—it became a *technological civilization*.

It is in this century that we developed the ability to move people and things in mere hours from any point of the globe to any other. This century's electronics can keep the population of the entire earth in instantaneous communication. We feed, transport, inform, entertain, clothe, heal, and kill each other by technological means that did not exist in previous centuries. Whatever we are engaged in doing for our survival, comfort, pleasure, or security rests on highly technological means. This is not a description applicable to the advanced nations alone. The underdeveloped nations are feverishly striving to become technological. Every nation of the globe now perceives its economic well-being, social stability, and national security to be dependent on the status of its technology.

This may properly be named the century of technology because it is the first century in which we began to understand that technological advance is not merely a source of benefits but can bring disbenefits as well and is a force with a powerful social impact. It is not that in earlier times social problems arising from technological change failed to surface. One hundred years ago and earlier, thinkers analyzing the developing industrial revolution warned of disruptions to the cultural structure to be expected as the machine age blossomed. But the big social alterations compelled by technological advance arrived in the twentieth century. The pouring of the population into the cities, the buildup of giant corporations, the unionization of labor, mass production, high standards of living, world wars, growing dependence of national strength on technological stature, the contest between free enterprise and governmental control—these and many other trends had made earlier appearances, but their real momentum came in the 1900s.

Society never completed its adjustments to these changes, so unfinished business remains, compounded daily by new dislocations. During our grandparents' era and a century before, scientific research and technological development were viewed almost universally as benefits to humankind. Society felt safe in avidly seeking to apply science and technology to enhance all of life's satisfactions. The one of my grandfathers I came to know worshipped science and technological advance (although his concept of technology was limited to the inventions of such as Thomas Edison, who was his idol). He died when I was still a boy, but I recall well his enthusiasm for the technology-based revolution he saw as under way. He was a philosopher of sorts and I was a bit precocious, but I don't recall our spending any of our one-on-one communication sessions expressing regrets about the potential detriments technological advance might bring to society.

If he were to descend for a visit today, I would have great difficulty explaining to him our inability to adjust to the technological breakthroughs steadily confronting us. I am sure he would cite the electric light, the telephone, the phonograph, and the automobile that came forth to enrich life during his period on earth and lecture me on the desirability of these inventions. He would emphasize that they made life easier and more pleasurable and that no problems existed in the way his generation handled their arrival. He would recall how he would look forward each year to more products and more factories in which to build them, which meant more jobs and higher pay. He would have difficulty understanding how we could do anything but welcome the additional technological advance since his time. He would look at our farm machinery and chemicals and marvel at how we have learned to produce much more food with less human effort from a piece of land, what with our chemical pesticides to stop the bugs from ruining the crops and our technological mechanisms for preparing the earth, sowing the seeds, fertilizing, and harvesting the crops. It was deemed marvelous in his time when ice delivery direct to the kitchens was begun, making it possible to preserve meat and vegetables. He would contrast this with our electric refrigeration, frozen foods, and microwave ovens to defrost and cook those foods.

He once explained to me how a sewing machine works and how the invention made clothing cheaper—not better, but cheaper. He would probably point now to pantyhose, fabricated by machine from synthetic silk invented with even more science, the fibers automatically produced. I can readily imagine his asking, with the never totally repressed mischievousness his family came to expect, "If my generation could adjust to the coming of sewing machines, why is it yours cannot surmount the breakthrough in pantyhose?"

Technology—To Us or Because of Us?

In earlier generations, technological applications—trains, telephones, automobiles, radio broadcasting—seemed to be occurring because of us. We the people were the source of the ideas; we knew they neither fell from heaven nor rose from hell by themselves. The process of implementing them was not perceived as out of control because the system apparently made the right things happen. Free enterprise brought technology and financial resources together, and when capital put at risk married technological advance, the offspring were new products welcomed by society. If attractive and sensible advances were offered on the market, the consumers bought them; if instead it turned out that a product was a mistake, then the market rejected it. For some important developments like bridges and subways, it was our democracy's elected government, rather than the free market, that determined what would be nurtured. All in all, what happened technologically in our society seemed to be in response to our wishes, whether expressed through the marketplace or through our government of, by, and for the people.

Now, however, it is a serious question whether technological advance has been happening *because of* us or *to* us. Let me illustrate.

What has that powerful communications technology, television, turned out to be? And how did it come to have such power? In the beginning, society saw the new invention as presenting us with a promising entertainment potential. Viewed this way, TV is a great success. An industry contributing greatly to the economy, TV involves many tens of billions of dollars annually in revenues and has provided over a million jobs. It has given communications more dimensions than could have been dreamed of when TV first was being developed.

But now we know TV is much more than entertainment. Television dominates the election of the president of the United States. TV has become so powerful as a communications medium that no one can be a credible candidate for the presidency, no matter how capable or experienced, who is not an outstanding television performer. TV's power is conspicuous also in the education of the nation's youth. It has so usurped the attention of our children as to be as influential in forming their minds as parents and schools are, if not more so. But TV's impact on determining our political leadership and on educating our young was not foreseen when television was first technically feasible. The citizenry did not debate ahead of time the desirability of the social alterations that automatically followed the development of TV. They just happened. They happened *to* us, not *because of* us.

Of course, TV does nothing that we do not foster by acquiring the sets and buying the products advertised. We have always had the option of stopping TV cold if the results are not to our liking. We can simply quit watching. We can junk the boxes or ignore the products pressed on us when the box is turned on. However, for the nation at large, such rejections are fanciful alternatives. TV is here to stay. We have to grant its permanent and pervasive presence and make do with limited attempts at manipulating its pluses and minuses. We can influence TV by our market response to it, and we can lobby for more control of TV programming. However, TV's impact transcends such influences. The power of TV is inherent in the very nature of the technology itself.

Gods or Devils?

If the example of television is typical, it is as though technological advance were an independent force in the universe, a phenomenon with a mission and strength not totally subservient to the will of the people that unleashed it. This is surely an exaggeration, but it is not an exaggeration to assert that advancing technology has become a potent force in altering our society. Moreover, we do not always know ahead of time all that the bargains we make, as we seek benefits from technology, may entail.

Either through spending our money in the free market or through actions our elected representatives in government take, mainly and properly to please their constituencies, specific technological developments occur. In initiating these, we may believe we are arranging for added conveniences, greater security, enhanced energy supply, more satisfying entertainment, travel at high speed with greater comfort, wider access to information, advances in medicine, more variety and nutrition in foods—altogether, a bevy of desirable favors the good fairy might bestow on us. But it turns out the fairy we are dealing with is often something else in disguise. From under the beautiful hair, perhaps actually a wig, and from behind the kindly countenance, maybe a mask, a devil may emerge later to slap our faces hard.

In America we worshipped science and technology in the past as idols or gods. We set these tools of our hard-working and inventive citizenry on a pedestal. This reverence seemed totally deserved, considering the blessings obtained through science and technology and the promise of prolific gains to come. We researched and innovated so well in areas of food and nutrition that we are able to produce sustenance well beyond our own national needs and thus to supply much of a world population that would otherwise go hungry. From the first horseless carriage we went quickly to ever more convenient and versatile automobiles, and then to airplanes; we added air conditioners and radios in cars and then movies in the sky. We can dial to connect automatically with most of the telephones in the world. In less than a century we have so automated production of the things we need that we have cut weekly work hours by half while producing several times more per worker-hour. We have conquered important diseases and provided numerous aids to the medical diagnostician and the surgeon in containing many other ailments. We have created cities full of structures in which the weather is controlled. We have put to our use every kind of matter appearing on earth, and we have synthesized a host of new materials not found in nature. We have visited the moon in person and sent instruments to take close-up looks at the sun's other planets.

With these accomplishments, it is understandable that we almost swallowed the myth that science could do anything, or at least came to believe that scientific research and technological innovation would continue to supply us with an unending stream of benefits. Some day, we expected, we would eliminate all disease, double the life span, provide for every material need with trivial human effort, and devote our lives to the most satisfying of pursuits rather than to boring, demeaning, or dangerous tasks required simply to stay alive.

America, we were proud to feel, epitomized the technological society. Using science and technology to attain a better life was the American way. It was in splendid harmony with the American dream of upward advance. Our technological society seemed to enjoy a compatible, mutually reinforcing relationship with the idea of free enterprise—investing, taking risks, exploiting Yankee ingenuity, constantly creating innovative and useful things, the system successful because everyone gained from the novel and better next step. We knew perfectly well we did not have the answers to all our social ills. Recessions, wars, crime, and other problems could be expected. However, in this American culture we had developed, we felt one thing could be confidently relied on: More scientific research and technological development would improve our lives with certainty, year after year.

Something's Gone Wrong with the Technological Society

But as the 1960s shifted to the 1970s, the long period of firm belief in technological advance as a limitless source of benefits came to an end. The conviction is now widespread that we became a technological society in too much of a hurry, without enough time to think, plan, select, approve, and adjust. Today many of our most thoughtful people perceive the technological society as not working, with more than just a little ailing it. It is seen as far less satisfactory than it ought to be, than what we had expected, and than earlier societies are considered to have been. As we move along in the 1980s, the technological society has crowded us into highly industrialized cities where too few can afford decent housing. Owning an automobile, instead of being a pleasure, is rather an annoying, absolute requirement for getting around, and we are required to allocate a substantial portion of our day to the misery of traffic. Moreover, we now see the automobile as dangerously polluting the air we breathe, and, from collision statistics, we know it to be an instrument of death and disability more punishing than any war in which we have engaged. Inflation, high interest rates, and high unemployment have moved in and show few signs of departure; they translate to our not being able to afford advancing technology's cures for our ills. Thus, new automobiles have been developed that consume less fuel, are safer, and pollute less, but our budgets require most of us to go on driving the old ones.

Technology, our century has proved, produces a plethora of detriments, as well as benefits, to society. It now seems that almost everything we do or use has the potential of increasing the cancer rate. Pesticides, fertilizers, and preservatives expand our food supply, but careless use of them subjects us to frightening risks. Despite tremendous effort, science has not yet found the universal cure for cancer; this is enough in itself to prove that science cannot do everything, not even eliminate all the dangers from technological advance.

The discovery of recombinant DNA may lead to mass synthetic production of interferon (a natural substance produced by the body to fight tumors and viruses and a possible cure for cancer), but it also opens new frontiers for biological warfare. Computers can make possible much more efficient handling of the information essential to operating our complex society. But at the same time, apprehension is rife about the coming computerized civilization because we fear it might lead to domination of our lives by computers programmed to schedule all production and distribution with little human selection and initiative. Are we going to become a society where people must respond to computerized signals and taped messages? In such a life it may be difficult to distinguish the role of a person from that of a semiconductor chip or a gear, as all mesh synchronously to achieve a precisely calculated, dictated efficiency.

Our technological society is not producing a job for everybody. Nor are inventions flowing rapidly enough to increase productivity so as to buck inflation. We have not discovered a novel, cheap source of energy. We are slow in developing needed substitutes for resources in increasingly short supply. Advances in medicine seem to go hand in hand with increased costs of medical service and hospitalization, so a financial crisis in our health system is pending. The pill is a welcome aid to family planning, but it has made for loose sex by our young and threatens the institution of the family. Our large cities are going broke, and the crime in them is escalating.

All the defects, devilishness, and doubts about our highly technological society seem to have arisen suddenly and together. At one and the same time, we discover that science is not an all-powerful, beneficent god; technological developments put in place to yield benefits also throw off unanticipated disbenefits; to prevent unacceptable harm we have had to create a huge regulatory bureaucracy; living satisfactions are falling, not rising; we are running out of natural resources; a technological war may end civilization. We can sympathize with someone who might say that if runaway technological advance is synonymous with the American way, then America may be a pawn of the devil and in urgent need of a new cultural pattern.

Finally, to confound our dismay, not only is high technology seen as a doubtful route to the better life, it is no longer even a symbol of world leadership by America. We used to be first in the world in technological developments. Technology and living standards seemingly advanced together as natural partners in defining the characteristics of our country, each giving evidence of our superiority. But now Japan, Germany, and others have taken over world leadership in important technological products, winning even in the U.S. domestic market. For many decades we led the world in productivity growth. Now all the other industrial nations are increasing productivity more rapidly than we are. We always granted that our principal potential military foe, the populous Soviet Union, could put more soldiers under arms, but we counted ourselves as ahead in military technology. Now they have more than caught up in many key areas. In their land-based intercontinental ballistic missiles, their warheads are more powerful. Superiority has almost lost its significance in our planning of military strategy.

The Triangle of Society-Technology-Liberty

We ourselves are to blame if our technological developments have gotten out of hand. If the results are ill-suited to the preferred way for humans, then we must humanize the technological society, organize so that we lead technological advance, not follow it. We must be masters, not slaves, of our technology. If we want things to be different in the future, we shall have to innovate. More technological innovation surely will occur, but that is not the critically needed innovation. For overall success in instituting change, the critical requirement is for social innovation, organizational innovation, innovation in the way in which we make use of science and technology, in how we select our objectives and priorities, in the pattern of our decision making.

To make innovations in running our society, it would be helpful to have some guidelines, an appropriate philosophy, a few main principles. One prescript suggests itself the moment we speak of needing to improve organization and arrange better means for selection of societal aims. Are these not functions we must entrust in great part to our government? Is it the government's role that needs changing? Is altering that the heart of the matter? This can be considered a reasonable view because surely what brings a democratic civilization into being, and makes it different from a mere collection of individuals all doing as they please, is that those individuals have chosen to give up some liberties. They have agreed to cooperate and behave in accordance with some restrictions that will apply to everyone, and they have created a government to lead in accordance with the citizens' desires, to issue edicts, and to police society.

The members of the civilization thus created would appear to have two guiding foci for their thinking. One focus, *liberty*, connotes the innate desire of human beings to be free, to decide their best interests independently and act as they choose. The other focus, *society*, implies a body of rules, accepted ethics, and patterns of organized approaches to common problems. These two conceptual centers, *society* and *liberty*, control civilization together. They reflect very different and essentially opposed needs, but they are connected. For a democratic civilization to exist, possess order and stability, and result in progress and satisfaction, the line linking society and liberty must be well traveled. The responsibilities emanating from the focus on society and the freedoms extending out from the focus on liberty bounce against each other and set boundaries along that connecting line. Civilization works when the influences of the two foci are in balance, when the liberty we give up and the regulation we impose on ourselves are acceptable, when the loss of some freedoms appears to be a price worth paying because civilization's preservation is deemed more important.

If we are going to improve our lives and the long-term future of the human race, we will do so only as an organized society, through cooperative action, as an ensemble of people agreeing to act in concert to make plans and policy decisions, to choose goals and implement them, to place this objective first and that one second. Such action will involve a continual contest between individual freedom and government control. Improving the technological society certainly includes refining the description of precisely how much we should ask the government to take on. As individuals we can be determined to better society and indulge in every liberty granted us to invent ideas and persuade others of their merits. However, each of us cannot create and put into effect a separate plan for America or the world.

Actually doing something together equates with government action, even if it is only our legislative bodies responding to public opinion. If we really believe all has not been going as well as it could, then we must also believe something is off balance between our individual freedoms and the responsibilities we have handed the government. If the United States is to follow a strategy for peace or economic strength or for curing disease or preserving the environment, the government will have to lay that strategy out. Actions of all kinds will be involved—organizational changes, new rules, bills passed to allocate funds. What occurs will have to meet the political tests for citizen acceptance, but the government will have to accomplish the task of integration.

But the two-foci picture, the concept of society with its rules and practices connected and contesting with the concept of liberty for the individual, is too simple. To encompass the workings of the technological society, we need to add a third focus: technology itself. The three foci, society-technology-liberty, can be visualized as the three vertices of a triangle, each vertex linked to the other two. From its vertex, *technology* connects with *society* and *liberty*, extending its influence along the lines joining it to the other two foci and in turn being influenced by them. The hard core of the technology vertex is scientific discovery—not totally independent, because it is connected to society for its sponsorship, and to liberty of thought and action for its pursuit—but independent in that what comes out are the laws of the universe which society and liberty cannot change and, once uncovered, must be accepted.

It is from this technology vertex of the triangle that a law of nature emerged in this century, disclosing that it is possible to convert mass to energy. This scientific discovery amplified the destructive powers of homo sapiens by a million times, a fact which civilization is powerless to reject or repeal and with which we now are forced to live. But also emanating from this vertex is an array of numerous, detailed technological advances that move out along the connecting lines to impinge on society, requiring adjustment in its rules and order, and on liberty, affecting the nature and amount of freedom each individual can be permitted. In turn, both the organizational patterns coming out of the society vertex and the pressures to preserve freedom radiating from the liberty vertex influence technological advance, determining which advances are fostered and controlling the nature and degree of application of technology.

The triangle of society-technology-liberty applies to little as well as big issues, to small communities and to the whole world order. Thus we give up the liberty to pass a red light at a traffic intersection in the interests of safe, orderly traffic flow in the community. We allow individuals to own guns but not A-bombs. We accept organized procedures for controlling the introduction of pharmaceutical drugs. From the technological focus information emerges that identifies smoking as a principal source of cancer; however, the workings of the triangle in this respect cause the society vertex merely to issue warnings, because the liberty vertex insists on the freedom of individuals to smoke. On a national scale, the liberty focus applies to groups of people and to corporations as well as to individuals. From the society focus arise rules that govern nuclear energy installations, determine government budgets for university research, set cable TV practices, and limit the liberty of automobile companies as to the kinds of cars they produce.

When we move to the international arena, the liberty focus is no longer a representation of the freedom of individuals and corporations. It is now representative of the liberty of individual nations to act unilaterally. The triangle of society-technology-liberty still applies: the nations of the world give up in some measure their privileges to act as they wish and opt instead for cooperation to obtain the benefits it brings. Thus, with space satellites making possible improved international TV broadcasting, nations have accepted procedures for allocating the limited radio spectrum and for assigning positions for satellites in space.

The triangle correctly pictures the situation even for matters of the security of nations, the sensitive balancing of the potential for peace and war. The countries of the world are not about to give up autonomy to a world organization to which their wills are subservient and which has the power to enforce compliance. Nations, from time to time, do form mutual security pacts, but each insists on having the liberty to go to war or to take any other action it deems necessary to safeguard its perceived ultimate national security.

Advances in nuclear science, coming from the technology vertex of the triangle, have presented to us so perilous a force of destruction that an unprecedented question has been raised. Has it not now become mandatory, if civilization is to survive, that the nations of the world deny themselves the liberty to destroy it? This matter is without doubt the most urgent and important illustration of the society-technology-liberty triangle of connected relationships with which this book is concerned. We shall take up this example first.

Chapter Two Nuclear War— The Definitive Example

Impelled by desire for conquest or urgent need for survival, human tribes have always fought each other and used their experience and imagination to improve their techniques of warfare. Are atom bombs another incremental step in the steady rise throughout history of fighting power and efficiency? Are they just the most recent addition to a list that has gone from clubs, spears, and bows and arrows to guns, tanks, and airplanes dropping TNT? Before nuclear weapons, the status of military technology at any moment in the annals of the human species was merely one factor, only at times the dominant one, affecting how wars were prepared for, what resources were used, and which strategies were most advantageous. In the past, wars commenced when some groups conceived it as advantageous and halted when the indicated further punishment was seen as too severe compared with possible gain. Nations clashed and bled each other, with some ending in ruin, the remnants of their civilization turned over to others to manage and use. Have all these patterns been changed by the arrival of the nuclear dimension?

It is easy to appreciate why some think the present nuclear arms confrontation is simply a continuation of past trends. Previous wars have killed tens of millions of people in a few years; nuclear warfare might result in a hundred million lost in half an hour. Still, with several billions now populating the earth, plenty of people would remain alive to start rebuilding the world society along the lines that existed before. Wherever humankind might have been headed when the holocaust began, it might lose a century or two in getting there, so the population numbers and general status of society in the year 2200 might turn out to be what, without the nuclear war, they might have reached in the year 2100. Those who see such a setback as not fundamental would argue that time, as the earth spins, is limitless. No requirement exists for the human race to reach any particular milestone at any given moment, year, or century.

Looked at in this way, the technology used in the continuing warfare of the species, the fraction of the earth's people killed off per war, the horrors suffered for specific lengths of time by the survivors—these are items interesting only to those later tracing the history of unintelligent life on earth. They do not make the history of the twentieth century, the century of arrival of nuclear capability, basically different from what occurred in the past and will probably go on happening in the future.

But such a view is wrong. The United States and the Soviet Union are committed to nuclear weapons programs which are beyond precedent as to peril, cost, and ultimate lack of any benefit. What we have developed now is not merely a higher level of destruction through higher technology. Both the United States and the Soviet Union have the capability of annihilating the bulk of each other's population, facilities, and infrastructure, poisoning much of all the earth's remaining surface, demolishing supply sources of water, food, energy, medical care, communications, and transportation, destroying the fabric of the social-political structure, and creating a chaos far beyond a short-duration Dark Age. The command and control of atom bombs presents such a complex dilemma that warfare might be triggered accidentally or irrationally or through panic or misunderstanding. And once unleashed, that warfare might not be limited, but rather be discharged full force.

This full force originates in a fundamentally new scientific principle, one unknown in previous centuries, that mass can be converted to energy. In the bomb dropped on Hiroshima, 1 gram (one-thirtieth of an ounce) of a certain mass was changed into a quantity of released energy equivalent to that obtained from exploding over 20,000 tons of TNT. It ended life for 100,000 people in a few seconds. The Soviet Union and the United States are each known to possess not mere grams or ounces or pounds, but hundreds of thousands of pounds of weapons-grade uranium and plutonium, which in the bomb reaction converts to the energy equivalent of billions of tons of TNT. Both nations have several thousand bombs, each 100 to 1,000 times more powerful than those used on Nagasaki and Hiroshima.

Most of the people of each superpower are to be found in 200 cities, and one bomb of a few megatons dropped on most such urban areas would kill or injure the bulk of their populations. Aside from immediate destructive effects, the ground blast of a single megaton bomb would create deadly fallout over a 1,000-square-mile area. Ten thousand megatons, presently available for release, would knock out 70 percent of the ozone layer that makes life possible on earth by shielding lethal levels of the sun's ultraviolet rays. It is estimated it would take thirty years for the protective layer to re-form. This kind of war cannot reward the winner because no winner would emerge. To institute this level of annihilation, with no benefit expected by the first striker, would be an act fundamentally new to civilization.

Most Expensive, Most Perilous, Most Pointless

The cost that has been sustained by each superpower to create this nuclear weapons capability, from the first experimental steps to today's full-scale implementations, is around a trillion of 1980 dollars, counting everything: research, development, laboratories, bombs, delivery systems, command and communications systems, support, training, maintenance, and numerous facilities and personnel costs. During the remainder of this century, the expenditures each nation seems compelled to make because of fear of the other's constantly expanding strength will surpass another trillion.

Such spending is enough to make the difference between failure and success in both nations' battles to cure their economic ills. The human and physical resources—the scientists, engineers, trained technicians, mines, factories, computers, materials, energy—needed to support the nuclear arms race, if invested otherwise, could increase productivity and lower costs to help curb inflation, create jobs to manufacture new and advantageous civilian products, uplift the disadvantaged by training programs, improve urban transportation, develop means of production with less impairment of the environment, invent substitutes for materials in short supply, raise world living standards, fight disease, and probe the universe to discover valuable secrets of nature.

With our liberty perceived as in jeopardy, we have believed we had no choice but to allocate these resources to the arms contest. However, no longer can either the United States or the Soviet Union hope to realize any military or political advantage or to enhance its security by the threat, or the actual execution, of a nuclear first strike against the other. The inevitability of severe retaliation precludes that hope. This means that the sole remaining value of a nuclear capability at this time is to deter the other side from using theirs. This applies to the European theater as well as to nuclear warfare on the two powers' homelands.

Of course, some in a position of influence today will not agree with this claim or, if believing it true, will be extremely hesitant to shout it out and reluctant to whisper it. Theoretical scenarios, in fact, are still being invented in an attempt to prove the assertion wrong, although the rationale put forth for the imagined circumstances in which a nuclear exchange (either limited or escalating to all-out nuclear war) would result in gain for the initiator is growing more absurd with each such recital. Starting a nuclear war with the object of winning seems an idea whose time, if it ever came, has passed.

For a substantial period of time following World War II, the United States alone possessed nuclear weapons with demonstrably enormous destructive power and credible means for reliable delivery anywhere in the world. We profited by its being understood by potential opponents that if hostilities were to commence between us, and we developed any doubts about our prevailing, we might employ the atom bomb. Through our nuclear weapons monopoly, we could intimidate others and limit their evil ambitions.

That era ended decades ago. When the Russians acquired a nuclear force, atom bombs dropped on them meant similar ones would fall on us. During the transition period of Russia's nuclear buildup, the Soviet Union possessed fewer ready weapons available for response than we; however this factor became increasingly less influential in setting our strategy. To cancel the seriousness of our threat to strike first, their surviving retaliatory strength only had to be enough to inflict on us an unacceptable level of extermination of our human and physical assets.

A Nuclear War in Europe?

For decades NATO's announced plans have included the possibility of our using nuclear weapons to halt a Soviet invasion of western Europe. While employing nuclear bombs has been put forth as a firm strategy to help us win if war should come, this option has been especially intended to deter the Soviet Union from ever reaching a decision to start a war. Along with having to contemplate that moving their armies into western Europe might be a bad idea for a variety of reasons even with our atom bombs not in the picture, the Russians have been forced by our nuclear strategy to include the possibility that early success in their offensive might find them involved in a nuclear exchange with the western alliance. This, they have been required to consider all these years, could have disastrous consequences for them, regardless of the amount of parallel harm they might do to us.

Many experts on our side who have continually analyzed NATO military strategy insist that our nuclear capability has indeed acted as a decisive deterrent. Of course, no one really can say with certainty whether the Soviet Union might have gone ahead and tried to take over western Europe had no nuclear strike options been included by NATO. Maybe the Russians did consider an invasion from time to time, and their fear that we might initiate a nuclear first strike against their forces was enough to cause them to hold back. Maybe not.

Either way, the situation is different now. Our dropping nuclear bombs on Soviet armies (say, in West Germany or any part of western Europe, or in East Germany or any part of eastern Europe) undoubtedly would be followed by their retaliating with their nuclear weapons. No matter what kind of nuclear strikes are employed first by either side, the other will not be powerless to retaliate with nukes. Western Europeans are well aware of this, and indications are growing that most want no part of it. It is one thing for western Europe's voters to embrace mutual defense plans with us that include sobering up the hawks of the Soviet Union with nuclear threats. It is another for them to accept the reality of nuclear bombs, whether of American or Russian make, actually detonating around their families. Yet we are still talking as if we might employ nuclear weapons should the Soviet Union suddenly dispatch tanks and soldiers westward.

Just as it was loosely accepted in western Europe in the past that NATO might use nuclear weapons, if necessary, to stop a Soviet thrust that appeared headed for victory, it is now becoming recognized, if equally loosely, that our European allies hold the equivalent of a veto power over our being the first to employ nuclear weapons. Whether the war might escalate to nuclear attacks on the homelands of the United States and the Soviet Union is of less interest to western Europeans than whether nuclear weapons of any kind will be set off in their highly populous, industrialized territories. If the U.S.S.R. were to invade West Germany, it is foolish to assume less than their total commitment to victory. To think the Russians would not respond with nuclear weapons if we used them first and, instead, would call off their takeover effort and withdraw is more than optimistic. It is ridiculous.

If insane leadership of both the NATO and the Warsaw Pact forces is ruled out, certainty of retaliation ensures that neither side will decide on first use of nuclear weapons. A nuclear force remains necessary for each, but realistically it is only to guarantee that the other will not employ nuclear means.

Limited Nuclear War and the Command Dilemma

If the Soviet Union attacks western Europe and air and land forces of the two sides begin large-scale conventional military encounters, might not a nuclear strike take place by accident? In the confusion of frenzied warfare could not an order to release nuclear bombs be issued through misunderstanding of the acts and intent of the other side, human error, fright, confused communications, and various mix-ups in decision making, some of which could be labeled as temporary insanity? Here the sensible answer is yes, quite possibly, even if not probably.

Nuclear weapons present a terrible dilemma of command and control. On the one hand, because the consequences of nuclear detonations are so awesome, all decision power to initiate them should be restricted to the highest level of command. On the other hand, unless provision exists for substantial decentralization of nuclear employment authority during a war, then destroying the top-level decision-making group would be a simple way to prevent retaliation. An enemy contemplating a first strike must be kept aware by its opponent that a system is in place for decentralized command and control, to guarantee retaliation under even the most adverse circumstances. Otherwise, the credibility of inevitable retaliation would vanish, and deterrence would disappear with it.

Certainly in this atomic age no sergeant or corporal will have the authority in the heat of battle to grab and launch available nuclear weapons at the advancing enemy. However, it is conceivable that some designated, authority-equipped team at the intermediate-control level might panic together, misinterpret incoming information, or react incorrectly to a failure to communicate with the command echelon above, and release an atom bomb. If this were to happen, a limited nuclear response from the other side might follow at once. Next, the immediately evident, terrifying results might bring the highest commands on both sides into the act with the use of their full senses. Conceivably, then, the bombing and even the invasion might be halted as suddenly as the initial nuclear strikes had begun. Negotiations might start then and there, both adversaries shocked into wanting to avert what they might fear would otherwise lead to inevitable escalation and mutual suicide.

In this way, by mistake, a limited nuclear war could take place in Europe. President Reagan was not wrong when, early in his administration, in answering a reporter, he refused to rule out the possibility of a short, localized nuclear exchange in the event of a western European war. Naturally, the western Europeans do not rate an accidental beginning of nuclear warfare in their countries as any more acceptable than a calculated one.

But let us suppose for a moment that those who still see a deliberately planned and implemented (not accidental) limited war as possible are correct. In the scenario that they would have us take seriously, one side or the other calmly commences nuclear strikes with the idea of using the nuclear dimension to ensure victory in a war in the European theater. Thus the Soviet Union, in this imagined tale, invades western Europe and simultaneously launches massive nuclear strikes against NATO's main nuclear stockpiles and all other European military targets, such as airfields, missile sites, military personnel concentrations, and centers for communications, command, and control. They avoid direct bombing of the population. However, throughout West Germany and the rest of western Europe the really important military targets are located of necessity in high-density population belts. Thus such a strike might kill tens of millions outright by blast and third-degree burns, and make other millions critically ill. Massive fires would spread to cause further casualties, and electric power and water systems would be incapacitated. No medical help would be available for most. Large fractions of the land would have to be abandoned as too dangerous for humans. The highways would be radioactive, and no trucks would move on them. Panic and chaos would reign as people sought evacuation (to where?).

Some flaws in this scenario immediately suggest themselves. Would the Soviet Union march its army into this deadly mess to take over? Even if they planned to delay moving in for many weeks after western Europe had been nuclear-bombed, severe fallout would still cover their armies and eastern Europe. Moreover, they would know their full-scale attack could not be over all in one split second. During the period from the first detection by the Allies of Soviet missiles or planes to the complete detonation of all Soviet bombs, the U.S.S.R. would have to reckon on the possibility of retaliatory nuclear strikes directed from western Europe at Soviet targets.

Some Allied planes would already have been in the air with nuclear bombs and cruise missiles when the airfields were struck by the Russians.

Some U.S. ground-based missiles in Europe would not be hit and would get off. The ocean-based nuclear missiles of the Allies—NATO can call on over 1,000 nuclear warheads on missile-carrying submarines—would all be unaffected by the severity of the Soviet Union's first strike at NATO land targets. Of course, the communications systems of all of the Allied commands would be greatly impaired. Confusion might be everywhere, so retaliation of any kind might be ordered and executed with difficulty. But certain high-level command structures outside of mainland Europe (for example, in London and Washington) would be left untouched to direct NATO submarines to attack the U.S.S.R. Thus, before implementing this scenario, the Russians would have to be prepared to take the enormous risk that nuclear weapons might detonate on their military forces and even on their major cities. For the Russians to assume no retaliation appears to require of them a mad optimism.

Sometimes this postulated Soviet action is expanded by adding the feature that after their surprise nuclear blow on Europe, they would threaten to strike the United States directly. They would try to blackmail us into abandoning both western Europe and the idea of direct retaliation against the Soviet Union in order that we might gain the elimination of danger to the continental United States. They would say, "Keep your submarine launchers and your ICBMs out of this European war, America, or else you will receive immediately devastating nuclear strikes on your homeland."

The U.S.S.R., with its leaders in their right minds, would have to assess as very high the probability that we would not comply with such a suggestion. To give in, the American president would have to decide to be disloyal to our allies and to disregard American troops stationed in western Europe, who would already have been under nuclear attack. Furthermore, the president would have to have confidence in a quick war-prevention deal with the very nation that had just demonstrated its willingness to perform a sneak nuclear attack. More likely, the president would reason that if the Russians were allowed to get away with a knockout blow to Europe, they might well choose to repeat the act on a larger, intercontinental scale. Instead of assuming that the United States would passively yield after a first strike on western Europe, the Soviet Union should assume that we would immediately send our land-based long-range missiles to hit Soviet targets, that we would take no chances that our ICBMs might be destroyed in what could soon be a preemptive attack on the U.S. mainland.

Would the Russians really count on our capitulation? Would that not be a crazy gamble?

It does not matter which side would deliberately start the nuclear war in western Europe. A first strike would produce the same negatives and offer no positives to the instigator. The resulting chaos that might enable an attacker to come in afterward and take over would be too likely an uninhabitable chaos, even if no intercontinental escalation were to transpire.

Direct Nuclear War between the United States and the Soviet Union

Let us shift now from the European theater to the continental centers of the two superpowers. Is a limited nuclear exchange on the homelands possible? Consider the following sequence of imaginary happenings.

The Soviet Union, without warning, launches nuclear weapons at our land-based missiles, demolishing all in a single thrust. They carefully design this attack to hit only the missiles, located in rurally located silos, and to minimize casualties and avoid damage to cities, industry, communication networks, transportation systems, and all other American facilities. The White House having been given a wide berth, the President orders our retaliatory systems on firing alert on the first indications of the attack. However, in accordance with preplanned procedures designed to ensure that the United States would never start a nuclear war because of a false alarm, the President expects to respond only after receiving unimpeachable confirmation that the Soviet warheads have arrived and exploded.

At this moment, the hotline, which does not answer on the Soviet end during the 20- to 30-minute attack period, comes alive with the following message from the U.S.S.R.: "America remains unharmed. Only the ability to damage the Soviet Union has been impaired." The President interrupts this message to show his anger, declares that we are now in an all-out war, and hangs up—but not before he hears the warning that any counterstrike in the U.S.S.R. will mean immediate, massive blows against this country, in effect, instant death for tens of millions, injuries for many more, and the virtual destruction of all the infrastructure necessary to continue as a nation.

The President, in this scenario, now turns to his advisers and asks whether we still retain the capability to inflict essentially total damage on the Soviet Union despite the loss of nearly half our nuclear weapons in the surprise raid. He is assured that we do have that power, but that the hotline warning from the U.S.S.R. has a totally credible foundation. In essence, we could destroy the Soviet Union but not their ability to destroy us in return. They certainly would be expected to release additional weapons immediately after our retaliatory launchings start, even before any of our warheads could arrive to destroy theirs. An attack on them would thus merely constitute revenge, and its consequence would amount to suicide.

But another alternative must be faced. The Soviet Union, continuing to use the hotline to us but now also speaking openly to the entire world, magnanimously proposes that the sensible action is for the two nations now to come together, agree to ban further nuclear exchanges and totally eliminate nuclear weapons for all the future. Leaders of other countries, concerned over the dangers to them of large-scale nuclear detonations, immediately deluge the President with importunings to select this peacemaking alternative.

If the Soviet Union's leadership were to become convinced that the United States would never choose to retaliate, to punish the Soviet Union for a first strike and deny their bid for heightened world power status, and if that leadership were willing to accept the risk of mass self-destruction, then they might be tempted to make real this fictitious tale. Of course, it would be an enormous gamble for them, because they could not be sure of the nature of the president's response. A serious knock-out attempt at our 1,000 Minutemen ICBM force, with the greatest of care to avoid other targets, would require some 2,000 Russian warheads at the yield level of a megaton each or more. Congress's Office of Technology Assessment has estimated that the resulting American deaths would certainly reach the level of millions, possibly even exceed 30 million. With such American casualties an accomplished fact and with the United States still holding 50 percent of its strategic capability, could the U.S.S.R. count on our suing for peace rather then retaliating? They certainly could not be sure ahead of time of the real effectiveness of their own attack, the reliability of complex military equipment being far from perfect and the scale of the necessarily unrehearsed strike being so vast.

They could not even safely assume that the United States would wait for their raid to be completed. Our leadership might elect instead to arrange that most of their warheads detonate over empty holes by directing our missiles at the Soviet Union even as their attacking rockets were still in the skies. (As a deterrent step, to minimize the possibility that the Soviet Union would seriously consider trying this strike, the president could announce as U.S. policy, years ahead of time, a firm plan to fire 100 of our 1,000 Minutemen at the principal Soviet military, communications, power, transport, and industrial centers if ever we should sight a large array of Soviet missiles in the skies traveling on trajectories terminating at our missile bases. Warheads they sent to destroy those 100 Minutemen would explode over vacant silos. To avoid any concern that this procedure might lead to the unacceptable starting of a nuclear war by a computer-controlled, "hair-triggered" false alarm, each of these launch-on-warning missiles could include an open switch, closable only by a coded signal sent by the president after the sighted Russian missiles had really landed and their bombs had detonated; if the switches remained open, our missiles would self-destruct high in space and no harm would be done.) The penalty to the Russians for being overly optimistic about either the technical performance of their untested operation or our response would not be merely an impaired world image. It would be the annihilation of their country. Thus many sophisticated and loyal American citizens think it inconceivable that the Soviet leaders would ever assume such a gamble. Others, strangely, believe that they might.

However, if the Russians were to act as the scenario describes, what should the American president do? Might the Soviet Union start by launching only a token raid to establish world leadership in a test of wills? If, say, they were to attack a mere 10 percent of our land-based missiles, how then should the president reply? Should perhaps some intermediate level of retaliation be attempted, perhaps the destruction of one medium-size Soviet city, or two transportation or communication centers, or three power and water sources, simply to show we are not impotent in either our remaining military power or our will to take up the challenge? What would then happen?

There is something fantastically and comprehensively unreal about scenarios that postulate a series of limited strikes between the Soviet Union and the United States on each other's country, the conflict a succession of accepted blows, each bomb delivery deliberately and analytically assessed before it is responded to, each response the result of that assessment and a careful tactical selection from numerous remaining target choices, each participant in turn awaiting the next strike of the enemy and planning to act only when its inning comes up.

If the Soviet Union should hit even a single American missile silo, why should we expect anything but enormous danger to us in what the Soviet Union might next do? Once the U.S.S.R. performs one dastardly deed, are we to entertain the weird thought that moderation and sportsmanlike rules will guide their following steps? Suppose we hit a single one of their silos as our rejoinder. Then do they next strike a second of our silos? Why would the Soviet Union, observing our lone response missile in the sky, headed for one of their silos, allow it to arrive and destroy their expensive missile? Since they presumably would be planning to launch a missile at us when it is next their turn, why would they not fire theirs while ours is still on the way? Anticipating their action, why would we go after their missiles at all? That would simply ensure that those missiles would be quickly directed at us, and nothing would be accomplished by expending each of our missiles so targeted—except to ensure that punishment was received in return.

If the two nations are going to inflict nuclear warheads on one another, then they will each want to minimize the damage the other can do, because even a single bomb brings effects it would be unforgivable not to avoid if possible. But the command, communication, and control systems necessary to engage in a series of well-calculated, limited nuclear strikes are especially vulnerable. Communication satellites in space can be knocked out. Ground antennas are basically soft and no radio communications with the ground will work without them. Airplanes can be blown up if on the ground and shot down if in the air and thus are vulnerable links in communication systems between ground command posts and missile silos or submarines. Yet if the idea of a stretched-out series of individual nuclear blows is to be given credence, the physical and human structure for decision making must be capable of surviving the nuclear strikes already completed. Means must continue to exist for knowing what has happened, what is now going on, what the enemy might next do, and how much damage has been done to the enemy and to one's own country. But unfortunately the command and communication systems might be the very first targets. Even if some residual command and control survives a first nuclear blow intended to destroy it, it might retain only the ability to command the release of the remaining nuclear weapons before they too are destroyed.

When one country contemplates starting limited strikes, it must expect that the retaliation it will receive will not necessarily be a strike in kind, more or less equaling its original blow. Instead, the first attacker may be the recipient of a large fraction of the payload inventory of its opponent. If it is going to be hit that hard in retaliation, then it should certainly not leave much of its capability sitting around to be destroyed, with only limited means remaining for its next attempt to win the war it started in error.

A few properly designed nuclear blasts set off at the right altitude over the United States—or over the most populous area of the Soviet Union, its western region—would create what is known as an "EMP" (an electromagnetic pulse akin to a mammoth lightning stroke), which would have a good chance of blacking out communications for hours or days or, for some parts of the system, indefinitely until repaired or replaced. Even without undergoing an EMP, it would be technologically difficult for the United States or the U.S.S.R. to maintain totally reliable communication and data-processing capability for all parts of the huge nuclear weapons command structure each superpower has built up. For each country, the system stretches for thousands of miles, from land to ocean to air to space and from silos to presidents.

Published estimates of the number of people involved in the U.S. Strategic Command and Control System put it at over 25,000.

If the command system is hit by a nuclear attack, an order simply to launch a retaliatory strike might still get through, but a complicated directive that calls for highly detailed variations (programmed by computers quickly handling the massive data collection, processing, and dissemination required after each blow) is completely different. For this, the president and other leaders in the command structure would first have to be located and briefed in four or five minutes as to the developing options. Then they would have to select and decide from among these, again in minutes. This would be followed by sending a train of reprogramming messages to the entire system. The worldwide sensors, data processors, communication systems, the top staff, and the implementing organizational levels would all have to be in tight contact, second by second, while the bombs and EMPs were impairing the network. This boggles the minds not only of expert computer-communications engineers but of anyone with common sense.

The concept of a deliberate, nonaccidental, limited nuclear war is unsound. Of course, any war, including even an all-out nuclear exchange between the United States and the Soviet Union, is limited in the sense that it is bound to end at some point. However, to those who believe in its real-life occurrence, a limited nuclear war is defined as one which, started for any reason, halts before it has gone too far, certainly before it has escalated to the point of using up the bulk of the bomb inventory of the two superpowers. Even should this limitation work, what military advantage will accrue to the initiator—unless the other side surprisingly surrenders after the first step or two?

Clearly, the determining factor in our planning is not really the extent of damage a first strike might inflict on us. It is rather to recognize that the Soviet Union can elect to open up a war of will and resolve, a contest of values, of liberty versus death, of independence versus compliance, and that we need to know ahead of time where we stand in resoluteness. Even though the leadership of our nation is constantly examining secretly what our alternatives and responses should be, surely these matters now should be a topic of public discussion. In the past we have considered it impractical, undesirable—too dangerous—for our president to speak of our retaliatory options or even to hint at considering such awesome issues as whether mutual annihilation is a preferred alternative to surrender.

We do not have to tell the whole world ahead of time precisely what we would do in answer to every conceivable kind of attack on us. However, if we could bring ourselves to talk about these things, perhaps we could establish that we have adequate resolve to retaliate. This might succeed in convincing the Soviet Union that it would be madness to contemplate a first strike on us of whatever strength. Both nations then might be more strongly motivated to come together to determine how to remove from the face of the earth the perilous nuclear weapons that neither side can sanely contemplate employing first. For both the United States and the Soviet Union, that would be better than the continued danger of mutual suicide triggered through misunderstanding of wills or temporary irrationality.

The Monumental, All-Out Knockout Attempt

So much for limited nuclear wars. What about the really massive, surprise, all-out first strike? It is common in some American circles to speculate about the U.S.S.R. reaching such a level of superiority in nuclear strategic weapons that in an unanticipated monumental strike the Soviet Union could obliterate virtually all our retaliatory capability, land, sea, and air. Then they could accept our weak retort, one perhaps harming their nation no more than they were hurt in World War II. To endure this damage, but thenceforth to be in control of the world, might be regarded as a worthwhile trade by the Soviet leadership. It is not possible to predict now their view of the world situation at the moment such a raid would be launched.

How would they structure such an attack on us? What are the chances of its being successful?

First, they would knock out our communications, command, and control systems. They would destroy our warning radars, communications satellites, the White House, and the airfields in the vicinity of Washington, D.C., and any others where airplanes that carry our airborne command systems might take off or land. The president and other leaders might never make it to the National Emergency Airborne Command Post (called "Knee-cap," a modified Boeing 747) which would either not get off the ground or be destroyed in the air, as would be SAC's "Looking Glass," their constantly airborne emergency center. (We must note that several thousand synchronized megaton-level detonations in the atmosphere over the United States would create an overpressure sufficient to destroy all airplanes then in the sky over a million square miles.) The Soviet Union would arrange—perhaps with locally based saboteurs using smuggled-in ground-to-air missiles—to knock out certain airplane-based communication links needed for contact with our submarines and for emergency communication with our land-based missiles.

They would destroy as many military planes as possible that were involved in any aspect of our strategic retaliatory system. Our submarines' home bases would be hit. High-altitude nuclear blasts would be detonated to create EMPs to black out communications.

Some of these first events would be caused by missiles launched from Soviet submarines. Those submarines would come up close to our coasts for these launchings, so that the warning time between our possible first observance of the launched missiles and the resulting detonations would be less than five minutes. These activities would be time-coordinated with actions by saboteurs to destroy ground radio and TV transmitters, relay stations, telephone switching centers, and other vulnerable land-based pieces of the total national communication system. After this, our network to detect incoming warheads would be largely inoperative and our command and control systems would have minimum capability.

The Soviet Union's long-range missiles assigned to destroy our land-based ICBMs would be precisely phased in takeoff timing so that all would land at about the same time as the missiles of shorter flight duration coming from the submarines close in. The Soviet Union might accept that our warning system, before its erasure, might observe Soviet ICBMs taking off and heading toward us. The president and others might thus learn of the possibility of an all-out strike, and our retaliatory system would be alerted. However, the soon-to-follow impairment of communication, command, and control might keep our ICBMs from receiving launching orders.

In the months and weeks preceding this first strike, the Soviet Union would make an all-out effort to keep track of our submarines and to culminate that search-and-track operation by destroying our submarines during the first stage of their attack. Using large nuclear warheads generously over the ocean regions where they thought our submarines were likely to be, they would not need ultimate precision in determining where the subs really were. Simultaneously, they would seek to destroy our ability to communicate with our remaining submarines and hence our ability to order them to retaliate.

Meaning to remove the United States as a serious future military threat, the Soviet Union would assign a portion of their ample nuclear weapons supply to destroying the U.S. infrastructure, going beyond our communication systems to transportation and electric power systems, water supplies, food and medicine inventories, centers of military forces and command, and civilian government leadership down to outlying areas. This would be intended to halt all substantive activities, throw us into total confusion and panic, and limit our ability to pull ourselves together to execute retaliatory actions. If they did all this thoroughly, even without deliberately seeking to annihilate our population, they most assuredly would vaporize, burn, or crush to death some 50 to 100 million Americans. At least as many more would be blinded, otherwise injured, or made critically ill by radiation. The fires and fire storms in the cities and the secondary explosions from fuel and chemical storage and distribution facilities would add chaos and misery to the fallout radiation. A large fraction of the population, with dead, partially destroyed bodies, flames, and rubble all about them and lacking food, water, medical help, shelter, and useful guidance, would be in helpless despair.

We are describing here a momentous nuclear strike by the Soviet Union which would be beyond anything in history in the complexity of its operation and in the number of separate, specialized human participants and machines that would have to be highly synchronized for the project to succeed. Any experienced systems engineer would say that if the entire integrated action, including the missile launchings and nuclear detonations, could be rehearsed—and doing so is inherently, of course, a fantasy—then hundreds of complete run-throughs would be needed before the system could be expected to work well. But not even one full-scale dress rehearsal is possible.

Of course, individual segments of the operation could be practiced. The system's controls could be tested and each component could report in, replying to command signals from the main control center. However, even limited practice runs would keep the airwaves busy. That they were taking place would be unmistakable to the monitoring centers operated by the United States. Before the Russians could hope to destroy our system of watching what they were doing, they would again and again have to test, adjust, debug, and refine their ability to accomplish that destruction in order to make it reliable. Their intentions, accordingly, could not be kept wholly secret. We would not be caught completely unaware.

In each of their practice exercises, as well as in the final act, should they carry it out, some things surely would go wrong. Some valves might freeze up in the pumping systems fueling their missiles before takeoff, or a control fluid pipe might burst, or during countdown some transistors might fail, or gyroscopes here and there might stick, or an essential instrument might malfunction. Human errors of numerous kinds would be made. Some orders would be misunderstood, and others would not get through. Also, one key aspect of behavior would be critical: All the partial rehearsals would take place with everyone calm, all knowing an error would not be the end of the world. However, the real thing would require cool, matching actions by thousands while all would be under greater doomsday pressure than a team of humans has ever experienced before.

What deviation from perfection could the Soviet Union tolerate and still consider their raid a success? Take one example. Suppose they knocked out 900 of our 1,000 Minutemen but the other 100 survived and were launched at the major Soviet resources, largely in the region within 1,000 miles of Moscow. Even if in that retaliatory act we sought to minimize personnel casualties, the industrial-communications-transportation-population density of the Soviet Union is such that those 100 Minutemen could destroy 80 percent of the infrastructure important for the Soviet Union to operate as a nation, including the foundations of their military. A third of their population would be killed, and millions more would be put out of action.

The Soviet Union has created bomb shelters to which they could send their leadership, including the local supervisory and management personnel of a substantial number of their cities and other important centers. (It has been reported they are building facilities for 100,000 higher-echelon personnel.) Just before the launching of a strike, those leaders might be directed to hide. Underground they would escape our retaliation and avoid temporarily the poisoned external environment it would create. Nevertheless, the bulk of their population would either be dead or be unable to continue with their lives, and the U.S.S.R. military force would be decimated and fragmented. And when the political leaders emerged from their holes into the still dangerously radioactive environment, they would lack the means for directing the shreds of the nation that would remain.

To the gambling odds experts in the Kremlin, the situation would look even less promising as they contemplated our submarine-based nuclear missile force. These vessels, a substantial group on station at all times, carry thousands of strategic warheads. Even if the Russians were to destroy half of our submarines—and it is a very big ocean and no information has surfaced to suggest the Soviet Union has attained a breakthrough permitting them to find our quiet subs with confidence—a thousand targets in the Soviet Union still could be destroyed in retaliation. It has been estimated that if all the warheads of a single Poseidon submarine (described often in public documents as over 100) were optimally targeted for industrial damage, 30 million Russian casualties might result.

Perhaps the Soviet Union would be successful in making it difficult for our command structure to get signals through to direct our submarines to retaliate. But could the Soviets count on our having failed to provide for delegation of command to fire these missiles in event of a communication blackout? What if our submarines were to launch their missiles at the Soviet Union precisely because the officers on the subs, after failing to get expected response signals from home over a certain period of time, decided that the homeland must have been the victim of a staggering nuclear strike? (They might surface an instrument to take a reading of atmospheric radiation residues to confirm their suspicions.)

Similarly, what confidence would the Soviet Union have that our retaliation procedures precluded launching our land-based ICBMs during attack or even on warning? If our observation system were first destroyed, we might take that fact as a definitive notice that our silos would soon be hit. If our command system has been engineered to allow for it, some of the Russian missiles would land on empty silos while the American missiles that occupied them a few minutes earlier would be on their way to targets in the Soviet Union. Could the Soviet Union dare rule out this possibility, considering the penalty of being wrong?

Part of our B-52 bomber forces are always in the air and would be on their way when the strike came. Perhaps the Soviet Union could shoot down some of these bombers should they attempt to cross the Russian borders, but they should expect those planes to launch cruise missiles from a distance. If only a small fraction of these cruise missiles were to land on their targets, the damage would go beyond what the Soviet Union's planners could regard as acceptable.

Meanwhile, what about the U.S. nuclear weapons based in Europe? A Soviet attempt to knock out the United States would surely have to include plans to neutralize Europe in parallel. Finally, to complete this listing of the gigantic risks the Russians would take in attempting a knockout blow at the United States, consider that the most well-engineered raid intended to be confined to America and western Europe alone, when of the scale we have described, would produce enormous fallout that would be felt in most parts of the world, including back in the Soviet Union.

We possess, fortunately, no data resulting from a full-scale execution of a nuclear attack by the Soviet Union and a counterattack by the United States. Even with full use of all of today's available knowledge about nuclear bomb effects, reliability of the engineering apparatus involved, and the workings of the pertinent military forces, no one can predict with certainty how it would all work out. So we have to say it is conceivable that the Soviet Union could be successful in their strike, that is, that they might receive negligible retaliatory punishment from us while essentially putting us out of commission. If that happened, they would then own the world. It is much more probable, however, that their bold thrust would be at best a mixed success. That would equate with suicide for them.

Aside from military effects, consider the gamble for the Soviet Union on the political and social fronts in executing such an attack. Suppose they were very successful with their strike, taking out the United States as an influential nation and making western Europe a conquered empire, sustaining considerable damage themselves but ending up strong enough for local recovery to occur in the Soviet Union in a reasonable time. Even so, the overall effect of the strike and the sudden loss of America's enormous contribution to world requirements would dislocate the world economy well beyond the predicament of Japan and Europe after World War II. In 1945–1946 the United States could and did step in and provide plentiful aid. Who would come to the rescue after a Russian-won nuclear world war? Certainly not India, China, Africa, southeast Asia, or the Near-East. With Europe and the United States in shambles, with the Soviet Union nursing its substantial if tolerable destruction, with the world's food, transportation, distribution, and production at radically reduced levels, and with people almost everywhere in dire need, the problem of organizing and managing the recovery would be immense.

The Soviet leadership is certainly not highly underloaded today as it goes about trying to feed its people, manage its internal economy, handle the relationships between Moscow and the distant ethnic republics within the Union, and keep in tow the present eastern European satellite states. Nothing about the past and present suggests that in a broad postnuclear chaos the weak integration and control structure of the U.S.S.R. would have the capacity to contain the huge, complex disaster area that would now include western Europe. As the conqueror, what would they really have gained? How could the Soviet Union possibly rate the potential benefits as attractive when compared with the dangers of disastrous failure and the accompanying physical ruination of their country?

The U.S.S.R., unless its leadership goes mad, is not going to try this all-out, unlimited attack on America. It would be as senseless as starting a limited nuclear war.

The Growth of the Peril

But the perils of accidental nuclear strikes persist and are rising. The risk of nuclear exchanges commencing through misjudgments or errors grows as the number of bombs increases and the variety and complexity of systems for delivering them proliferate. The more times the nuclear bomb inventory of each nation can wipe out the other, the vastly greater the danger. The more dimensions of nuclear warfare there are—the greater the number of people, missiles, submarines, airplanes, satellites, computers, communication links and interactions involved; the more intricate the controls; the more widely spaced geographically the participating components—the more likely it is that something can go wrong during some world crisis.

The more thorough and complete the destruction potential, the more a nation responding to a nuclear strike is driven to measures not characterized by calm reasoning. For instance, if the enemy is seen as having the ability only to impair, but not totally obliterate, one's communications systems, then it is conceivable that time can be taken after receiving a nuclear blow to assess what has happened and to consider options carefully before settling on a retaliatory action. On the other hand, if it appears the enemy has the capability to obliterate one's command system, then it becomes urgent to order a retaliatory reply before being precluded from doing so by further enemy action. Pressure to act before being totally destroyed can become irresistible and can lead to preemptive moves, instant responses, irrationality. The human components of the decision-making apparatus become outmatched by the speed, power, and irrevocability inherent in nuclear weapons systems.

Simultaneously, costs escalate and not just for the weapon systems themselves. It is straightforward, even though increasingly expensive and dangerous, to do what both superpowers are now doing, namely, to keep producing more bombs and delivery systems, hoping enough will survive the enemy's first strike no matter how powerful—and that the would-be aggressor, realizing that, will be deterred. The expected strength of the surviving retaliatory force is a measure of the strength of the deterrent acting against the first strike. It is far less straightforward to design and put in place a survivable communications system for command and control because it is simply not natural to harden the components of such systems. To do so requires taking incredibly uneconomic and technologically impractical steps. Just as it is theoretically possible, but prohibitively expensive and impractical, to construct underground shelters to house 200 or 300 million people to ensure their survival in the face of the most severe knockout blow an enemy could muster, it is also too costly to arrange for the survival of the means to communicate with and control the personnel and equipment needed for retaliation after the strongest of surprise strikes.

As the dimensions of nuclear arms grow and the destructive power on each side rises, the situation worsens in other respects. Both sides may soon feel it necessary to launch programs in space-based weapon systems, perhaps including laser beams or nuclear-tipped guided missiles, ready to knock out the enemy's communication satellites which are necessary to command nuclear attacks or retaliations. The country owning those satellites similarly will set up space defense systems to protect them. In this way, space-based warning systems might turn into space-based war-making systems, attacking and defending against each other with ever greater complexity required in space and in coordination with the earth below. The whole will escalate in its economic cost and, in the end, in its uselessness to either side.

Innovating to Attain Real Nuclear Arms Reduction

We can get out of this increasingly penalizing nuclear weapons race only by innovation—not technological innovation, although science and technology will always be involved, but rather by social and political advance. Fortunately, to do things differently does not require changing the basic character of the human race. But certain of our defects, even if not fundamental and permanent, are of such nature that they cannot be altered quickly, not in decades, perhaps not in centuries. Thus we cannot expect to arrange soon that disputes between people in the world will never again be settled by the use of force. We cannot compel in the near future, through even the most brilliant of social inventions, the acceptance by all nations of a world government, a grand democracy with a single, central control. The way the world has gone in this century does not suggest a greater homogeneity of goals and a general equality in standard of living and per capita allocation of the world's resources among nations, or a decrease in the number of groups insisting that happiness and security can only be theirs if they are allowed to become independent nations. So wars remain likely.

But it is not impossible to innovate so as to lower the threat of a nuclear war. Large-scale nuclear weapons reduction by the United States and the Soviet Union will take place when two conditions come to exist. Both already have a strong base of credibility.

The first requisite is that the leadership of the two countries advance from today's beginning realization to complete conviction that in view of the strength and resolve of both sides, nuclear weapons no longer offer advantage to either. Not in their actual employment. Not in threat of their use. The heads of governments must be willling to state this unequivocally, and they must really believe it. The second requisite is that means must exist for each nation to be satisfied that the other has complied with the agreed-to arms reduction without requiring internal inspections of either country by the other.

Consider the first requisite, the settled conclusion that the nuclear race offers no military, security, or political gain to either side. Before his death

Brezhnev made emphatic public statements that a limited nuclear war was not possible, that neither side could win because the inevitable escalation would lead to the destruction of both societies, that the Soviet Union disavowed a first strike, and that any nation that contemplated starting a nuclear war must first have decided on suicide. If the Soviet Union has really accepted these ideas as basic facts, then they should be willing to negotiate real nuclear weapons reduction when we propose a practical way to satisfy the second requisite. On the other hand, if Brezhnev was solely propagandizing and the Kremlin's true views and plans are quite different, then we cannot succeed now in arranging mutual disarmament steps with them. We will learn which by entering into negotiations.

When both nations become completely convinced, and convincing, about their resoluteness to retaliate against a nuclear attack, and are perceived as possessing the means to do so even after the other has struck first, then one of the mandatory conditions will be met.

Now, as to the second requisite, how can the two nations substantially reduce their nuclear weapons, with unquestionable verification, yet without the need for (the practically unattainable) on-site inspection? The answer suggests itself when we take note of a problem the nations must solve when they dismantle nuclear warheads: the disposal of the resulting nuclear waste. What should be done with the fissionable material taken from the bombs? Let it pile up? Bury it in the sea? The answer is to transform the potentially nasty negative into a useful positive. The bomb materials should be converted to fuel, an economically and technologically sound operation, and used up in nuclear reactors to generate electric power.

Despite all the official secrecy, both nations know full well that each is capable of judging with adequate accuracy the bomb stockpile of the other and the weight of nuclear materials contained in the bombs. Suppose that, with neither side asked or expected to disclose its inventory totals, each agreed to bring a weight of fissionable material corresponding to substantial arms reduction—say, a third or half of the totals as a first step—to a jointly supervised facility where it would be reprocessed to a form ready to produce electric power in a conventional nuclear reactor. Calculations based upon published estimates of warhead stockpiles indicate that enough equivalent fuel is contained in the warheads to generate electric power for decades at a level much greater than any existing or planned nuclear power plant.

Negotiations would fail if either side was dissatisfied with the quantity of materials proposed for delivery by the other. For instance, the United States might offer a certain weight of plutonium or U-235, claiming that this represents a substantial reduction of our bomb inventory while disclosing no classified information about our total bomb yield or about the relation of destructive yield to weight of fissile matter for our bomb designs (as would be necessary to back up the claim fully). The Soviet Union, on the other hand, might propose a far more modest contribution of materials, perhaps maintaining they have a smaller stockpile than we or indicating that, owing to undisclosed advances in their warhead designs, they obtain high bomb yields with relatively little critical matter.

However, it is likely that those fundamentals of science and technology which know no national boundaries would exert a moderating influence. Neither nation would come to the conference with the idea of bamboozling the science and engineering experts of the other, because both would know ahead of time that such attempted fakery would not work. A loosely agreed-upon range of bomb reduction would get translated by negotiation into hard commitments to deliver detailed weights of specified fissionable materials.

Nuclear materials delivered to the controlled reception site no longer could remain simultaneously in the nations' bomb arsenals. With elegant simplicity, inspection and verification would be inherent, immediate, and automatic at that site.

Of course, the implementation of this approach to weapons reduction does not require the creation of new reactors.[1] The most straightforward alternative is to route the newly available fuel, under international safeguards, from the conversion facility to nuclear power plants anywhere, offering it at market price, with the partners sharing the net revenues.

What if, immediately after the agreements, one of the countries secretly embarked upon a program of rapid replacement of the contributed bomb materials? With a substantial fraction of the crucial innards of the present bomb inventory given up to the conversion facility, a very high rate of production of fissile matter would be needed to replace it quickly. If one side were to cheat in so big a way, the scale of its operations would cause the other side to learn what is happening through existing, off-site monitoring systems.

How about the distinction between the store of bombs and the effectiveness of the overall weapons systems, of which the nuclear warheads are a part but which also include the guided missiles that deliver the

---

[1] It is almost irresistible to imagine setting up a new electric generating facility on one of the islands off Alaska, adjacent also to northeast U.S.S.R. By use of two big cables, half of the generated electric power could be sent to each nation. Unfortunately, unless the nations decided to put huge aluminum-producing installations or equivalent industrial operations in the area, this suggestion may be a fanciful way to illustrate a point but is not a practicality.

warheads? As missile guidance accuracy advances, as new bomb designs are developed, and as the remaining nuclear materials are spread optimally over more warheads and more missiles, a reduced and fixed nuclear materials inventory conceivably could be restored to its preconference destructive power and the peril would not have been diminished. There is no lasting answer to this problem. We cannot by arms-reduction agreements halt all advances in military science and technology, and it is unrealistic to expect any negotiation today to eliminate all nuclear phenomena forever from the military forces of the world. However, a determined effort to reduce nuclear arms is justified and needed, if only to combat the trend toward a continuing augmentation of nuclear weapons.

What we have described would amount to a first step, lowering the dangers and costs of nuclear weapons systems while nevertheless retaining them, perilous and useless still. One nuclear weapon in the hands of the Soviet Union sufficient to wipe out, say, New York City or Los Angeles or Washington, and one we hold large enough to destroy Moscow, should together be enough to deter each side from using its nuclear capability. So extreme is the human damage from a 1-megaton bomb exploding on any large city that the presence of nuclear weapons in the arsenals of both nations would deter a first use by either, whether each possesses one weapon or ten thousand. If once accepted as true, this concept could lead to the attainment of nearly total nuclear disarmament in a series of steps.

Nuclear Weapons Proliferation

Other complications create limitations in what the foregoing approach to nuclear disarmament might accomplish in the short term. For instance, nuclear capabilities are possessed by nations other than the United States and the Soviet Union, and those nations also must be brought into the act. Furthermore, blackmail is possible by some nations which possess little interest in arms-reduction efforts but which might see an advantage in threatening to employ the few nuclear weapons they might assemble. Finally, the negotiation difficulties surrounding tactical nuclear weapons stationed in Europe are very real and are not separable from strategic arms negotiations.

But none of these issues would be worsened by the mutual nuclear-reduction approach described. On the contrary, such a program would do more than merely lower the probability of nuclear exchanges between the two superpowers. It would improve the overall world arms-control environment and so should militate towards nuclear weapons reduction everywhere. The pact would imply recognition of the futility, peril, and enormous costs of nuclear warfare, and this recognition would be bound to influence all other nations, including any now contemplating or actually secretly engaged in developing a nuclear weapons capability.

The United States and the Soviet Union, working together, would carry enough weight to influence strongly almost every other government's plans for nuclear weapons. It is worth emphasizing that anticipating living with nuclear blackmail from the less predictable, dictator-controlled nations should be enough in itself to drive the two top nations to action. Without cooperation by the United States and the Soviet Union to invent ways to control nuclear terrorist acts, such activity may surface in a very short time. In 1980 there were 400 nuclear reactors operating in 28 different countries. All were presumably created to produce peacetime electric energy, but inherent in the process of such generation is the incidental creation of plutonium within the fuel rods of the reactors. With the proper apparatus and personnel this plutonium can be extracted. Enough is already becoming available from present reactors to produce powerful nuclear bombs at the rate of one per day.

The head of the International Atomic Energy Agency of the UN recently complained that their 140 inspectors cannot begin to deal adequately with keeping track of the 15,000 tons of nuclear material from reactors housed in nearly 350 sites worldwide. Aside from the 115 signers of the 1970 nonproliferation treaty, we must be concerned about the nonsigners that the agency is not even invited to inspect. For instance, Pakistan, where negotiations to beef up inspection procedures are stalled, is putting finishing touches on a uranium enrichment plant and a nuclear materials reprocessing facility that they do not even intend to put under international safeguards.

Nations currently without nuclear bombs can obtain the know-how, materials, and equipment needed to produce nuclear weapons by buying them from the governments of the countries that possess them or from private citizens and companies. All of the latter should be controllable by their countries' governments. When the two most powerful nations engage in a potentially disastrous, economically punishing, and basically useless race to increase their enormous nuclear arms capabilities, then the meaning of being responsible becomes confused and the concept is diluted. All nations then find it easier to succumb to the temptation to profit from export of their nuclear know-how and equipment.

Removing the threat of nuclear terrorism is dependent on first establishing sensible leadership by the two superpowers. Here the triangle of society-technology-liberty displays for us a clear guide. For the foreseeable future the nations of the world will insist on retaining the liberty to take any step they perceive as necessary at any time to ensure their survival. But the United States and the Soviet Union do not require for their survival the liberty to commit suicide, the liberty to utterly destroy the other and, because of inevitable retaliation, the then-certain concurrent liberty to destroy themselves.

Technology has thrust the nuclear bomb into the triangle, and now social progress is required to contain it. That needed social advance will consist, as always, in creating and following innovative patterns of behavior, setting rules that limit the freedom of independent action of the participants in the civilization. When it is finally understood that the only liberty the nations need relinquish is the freedom to perform an insane act, then the society-technology-liberty triangle will become a stable figure and civilization will survive.

Chapter Three Pervasive Technology— Elusive Free Enterprise

If the world of the future were magically made immune from any threat of war and, accordingly, if technology were used only in peaceful pursuits, this would not necessarily free us from the danger of the misuse of technology. A society in which technology is employed carelessly would not be a happy society nor one in which human welfare is always enhanced. To attain that desired end requires at least some deliberate effort to relate technological implementation to articulated objectives.

The nuclear arms race is the most dangerous of all impacts of technological advance on civilization. However, when that menace is understood by the participants who have created it, then the action required to diminish the peril is straightforward, even if extremely difficult to arrange: We and the U.S.S.R. must engage in real, large-scale, mutual nuclear arms reduction. In contrast, the impact of many other kinds of technological advance on society, although less ominous, is also less clear, and it is less evident what courses of action should be followed to maximize the benefits and minimize the negatives from implementation of this technology.

What exactly makes it so difficult to arrange that science and technology be used fully to the advantage of the society? It is because science and technology are not phenomena independent of society, and because society's goals usually are not clearly set out. We cannot look over proposed scientific and technological projects, pick those corresponding to society's known objectives, isolate them, and go forward. Aside from pure research into the laws governing the universe, happenings on the scientific and technological front are not separable but are rather inextricable components of our societal problems and opportunities. Curbing inflation, attaining economic growth, avoiding war, ensuring adequate energy, protecting the environment, fighting disease and hunger, creating jobs—none of these can correctly be called a technological issue. But technological advance is heavily involved in them all. Social, economic, and political problems always intersect each other, and critical aspects of technology can be found perched right in the middle of the intersections.

Technology and Inflation

It can be argued that the fundamental issues of the world are economic and that all social, political, and war-peace problems are spawned by economic ills, scientific and technological aspects being secondary. Thus we note the president has a council of economic advisers but does not feel the need for a council of scientific advisers. He is probably briefed very frequently on ideas for bringing down interest rates but rarely on how the United States can win the international contest for technological superiority. However, technological advances can be powerful tools when fighting economic battles, and the status of our technology is basic to our economic strength.

For example, if American productivity could be raised suddenly through breakthroughs in manufacturing technology, i.e., if innovations enabled us to turn out more per worker for the same investment to equip that worker, then costs would drop, supplies would increase, inflation could be fought more easily and successfully, and our standard of living would rise. If our technology for designing and building automobiles were suddenly to take quantum leaps, American cars would beat out foreign-made ones in cost and performance and we could stop worrying about declining employment in our automobile industry. If we could generate cheap, plentiful, safe, and nonpolluting energy through scientific discoveries, then we would not have to send dollars overseas to buy petroleum and fuel prices would fall. If innovations in information-handling technology could be installed everywhere instantaneously, we could cut overhead costs in everything we do. If, by radical inventions, we could defend our nation readily and inexpensively against any action of a potential enemy, then national defense expenditures would come down. All these technological developments, if they occurred, would offer new opportunities for lowering taxes, making more funds available to develop additional novel products for the civilian market, and decreasing unemployment.

An important example of technology's involvement with economic problems is found in the phenomenon of inflation. A few people anticipated inflation years ago, borrowed heavily when long-term money could be obtained at low interest rates, and bought up assets when the prices were low. Today such smart investors (or lucky speculators) could sell out, pay the capital gains taxes, count their money, and celebrate inflation as a great bonanza. Their experiences provide no basis for understanding how inflation and high interest charges hurt most individuals, businesses, and governments. Inflation is known to be villainous, but it is not always recognized as restrictive of technological advance. Indeed, inflation is the most limiting of bottlenecks to our realization of the sweet fruits and solid nutrients that technology can offer society.

Most technological innovation arises in industry, where it is routine to invest capital and apply expertise to advance technology. In good times, the cash generated from operations is bountiful, and industry leaders can consider a broad range of new ideas to pursue. In periods of recession fewer funds are generated for risk taking. In inflationary times, we can have the appearance of satisfactory profits, while the flow of cash truly available for investment is disappointingly low.

The Internal Revenue Service's accounting rules for industry cause announced earnings during an inflationary period not to reflect the true costs of doing business. For instance, as manufactured articles leave the plant, the parts they contain can be rebought or remade for the next production run only at prices substantially higher than those originally paid. It often has been the practice to report as earnings the paper profit realized simply by having bought earlier, at lower than present prices, the materials and labor that went into the products sold. Companies might register earnings rather than losses only because of this accounting practice.

The Internal Revenue Service is uncharacteristically kind to a family that must change location and is forced to sell its present house and buy a new one elsewhere. A house bought many years before for $50,000 might bring $200,000 today. Without special dispensation, the government would take its income tax cut of the $150,000 profit on the sale, around $35,000, leaving the sellers now with cash of only $165,000. If they had to pay $200,000 at their new location for the equivalent of the house just sold, the family would have to go into debt for $35,000 to swing the deal. Their reported income for the year would be higher by $150,000, but they would suddenly owe $35,000. Fortunately, the IRS forgives this tax if the seller buys another house quickly. But it does not do the same for industry.

Industry's buildings, machines, laboratory equipment, and the like, all necessary to stay in business, wear out and must be replaced as time goes on. Accounting rules allow a fraction of *original* purchase prices to be listed as depreciation expenses, but with inflation, the true replacement prices are always much higher. So again, the earnings reports are inflated, and the government takes more of the cash being generated by company operations in income taxes, leaving less to be invested in technological advance and causing higher company debts even as interest rates go higher and interest expenses soar.

Typically, a company will have bought its physical assets years ago at half of today's prices. Thus, when technological corporations report the important ratio of return on assets employed (around 15 percent for the average stable technological corporation), both the numerator (earnings) and the denominator (assets) should be corrected for inflation. Then the real profitability would often come down to 5 percent or less, a miserable rate of financial return, considering the risk. During the past decade of inflation American industry has not been generating adequate cash flow to invest properly in superior facilities, to learn how to make existing products at less cost, to invent new products, and to advance productivity in general. Inflation has become the principal enemy of technological innovation.

All the while, technological advance has been a potentially powerful weapon with which to attack inflation. Here inflation is the dragon terrorizing the land, and technology is the brave knight poised to slay it. Of course, inflation really is too much money chasing too few goods, and we experience inflation when the money supply increases too rapidly compared with the goods supply. To stop inflation, we don't require technological advance. We need only to stop the overexpansion of the supply of money, and in theory, the Federal Reserve can do just that. But social-political considerations stand in the way of the straightforward application of this idea. Stemming inflation by curbing money supply growth while the government's budget is in a high deficit range can bring on a severe recession and high interest rates. Then the political pressure on the government to take steps to ease the resulting unemployment and bring interest rates down, even if inflation is thereby reinvigorated, can become irresistible.

If the government's economic policies are wrong, then technological advance cannot save the economy. However, in the real political world, what occurs is a lot of partly wrong and partly right government actions, and few economically perfect ones. While all the frenzied government measures are being taken, it would help greatly if technological advance could be employed with alacrity to heighten productivity, lower costs, and increase the supply of goods. Then the money supply could be allowed to grow proportionately, and inflation, high interest rates, and recession would be much less likely to result. Ironically, the very existence of inflation has made it harder to raise funds for investing in the technological progress that would counter inflation. That is, because we have inflation, we find it difficult to use technological advance to aid in halting inflation. Inflation and the government's less than effective fight against it have created an economic environment that decreases and discourages investment in new technology.

Over the last decade the fraction of our GNP (gross national product) that was invested, in comparison with what was consumed, was much less than this ratio for Japan, Germany, and other major industrial nations. Meanwhile, our production facilities have been growing obsolete. If the latest available technology were incorporated into the nation's production operations, efficiency and productivity would rise. However, the incentive for investment, the real financial return, has been too low for advancing technology to be employed fully to do its part in fighting inflation.

For example, low returns on investment and high interest rates have lowered the rate of founding new high-technology companies. Fewer were launched during recent inflationary years than like periods a decade or two earlier. It is significant that the following new ideas originated with individuals not involved with large corporations: atomic energy, computers, cellophane, DDT, FM radio, foam rubber, insulin, lasers, the Polaroid camera, radar, rockets, streptomycin, xerography, and the zipper. Most industrial research and development is done by the large technological companies. However, as we can see from this list, a diminishing of contributions from new small entities is bound to be harmful to the economy.

My departed grandfather, on a return inspection tour, would find it hard to understand that during our recent inflationary decade the public has been favoring consumption over savings. In his generation, the concept of personal savings was held high and debt was feared as a dangerous menace. But the prevailing religion in recent times has been that it is better to buy immediately, even if one has to borrow to do it, because whatever is purchased surely will be priced much higher later. This thinking has been further intensified by our taxing of savings. Suppose you realize 6 percent on your savings, the annual inflation rate is 12 percent, and you are in the 33 percent income tax bracket. You will get to keep only 4 of the 6 percent income on your savings, giving the IRS the other 2. This 4 percent less the 12 of inflation comes to a net loss of 8 percent. It is better to buy something right away that you believe you will need in the future.

Under the inflationary environment of the last decade, the technological industry has become conservative and cautious. What would you do if you were the manager of a company and you saw that while you were able to announce reasonably favorable earnings performance, your real return on the real value of the investment entrusted to you was discouragingly small and your interest expense on your heightened debt was rising? You would take the meager new investment funds truly available each year and, like your peers, bet them on safe incremental investments in the products or the plant. You would be aware that it was becoming riskier to take chances on big steps. If you could acquire extra cash, you would consider buying up another company that you thought was undervalued in market price compared with the replacement cost of its assets. You would not be likely to favor hurting your already poor earnings report by expenditures on long-range research and development. If you were to start a long-term speculative project, you might find it difficult, you would be aware, to obtain the funds needed each year to complete it.

The American habit is to rate the performance of business management by a conspicuous calling out of a simple score: earnings for the quarter just ended, which are then compared with the previous quarter and, more particularly, with the same quarter a year before. Operating managers usually receive an annual base salary (established roughly by the size of the responsibility compared with compensation for like-sized responsibility in similar businesses) plus a bonus set by the reported earnings for the just-completed year. Projects in advanced technology usually take longer to mature—from the original approval of an embryonic effort, through the loss-certain development and start-up phases, to a profit period—than the average length of time an operating executive remains in charge of a specific unit.[1]

---

[1] A study has shown that in recent years the average time the chief executive officer of a public company or the general manager of a corporation division stays in that job is between five and six years.

Thus managers have a bias against long-term projects where high start-up losses will hurt their financial remuneration and prestige and where the eventual success, if it should later occur, will add only to the standing and income of their successors.

Fortune magazine has quoted the head of a large private company, speaking of the contrast with the typical public company, as follows: "If I knew my compensation next year would be based on this year's return on equity—I wouldn't act the same. You've only got a few years at the top in a public company to make your killing. You want to put every penny on the bottom line to wind up with the juiciest retirement package you can get." And again, a former chief executive officer of one of the nation's largest corporations: "Every CEO says he plans for the long term. But every CEO lies. He's always temporizing with quarterly earnings. If he doesn't hack it quarter to quarter, he doesn't survive."

But if inflation and high interest rates enhance the tendency of management to be short-term-oriented, government officials are inherently more so. Here the pressure to concentrate on the immediate period ahead is enormous. Significant elections at the national level take place on a two-year basis. Even the presidency involves only four years (the first for getting acquainted with the tasks, the second and third to get something accomplished, and the fourth to run for reelection). Steps that might benefit the nation in later years get low priority if they appear politically harmful now.

Thus basic scientific research in universities is greatly handicapped by inflation. This foundation for our future depends largely upon government grants and, to a lesser extent, on industry's philanthropy. The government, when desperately engaged in the short term in trying to curb government spending, becomes a weaker source of university funding. Meanwhile, industry, suffering from low cash flows, recession, and high interest expenses, has less left over to give to universities.

As a final example of the way advancing technology is intertwined with inflation, consider the situation of America's cities. We need better transportation, waste disposal, environmental controls, crime prevention, health care, and schools to make the cities decent places to live and to enable them to operate efficiently. Improvements in all these areas are possible if the latest technology can be employed. For instance, we can move people and things about in our principal cities at lower cost and with less pollution, less energy dissipation, and fewer accidents than go with today's typical city transport, mainly consisting of private automobiles. However, little incorporation of technological advance to improve our cities appears affordable. During our inflationary decade, the financial condition of virtually every large city became precarious.

Inflation is not a necessary attribute of a technological society. Its presence is a symptom of defects in the political organization and management of the nation. However, no approach to curbing inflation can succeed with American voters in the last analysis if they do not see it as answering their insistent demand for a steadily increasing standard of living, still less if it requires their acceptance of a substantially decreased flow of goods and services through a severe, prolonged recession. Equally, the disadvantaged, those whose share of the nation's output is well below average, will not and should not give up their aspirations for a higher level of participation. This means the tools of science and technology must be kept sharp and used diligently because through their use comes the increased supply necessary to solve the political dilemma of inflation.

Ubiquitous Technology

Even if inflation were defeated, plenty of other important social-political-economic issues would remain, with technological advance inextricably immersed in every one, especially because our technological society begets more technology. We can illustrate this by noting some key aspects of three growing societal problems: ensuring the supply of energy and natural resources, controlling the hazards of technological implementations, and providing for the nation's health care.

Compared with the situation 100 years ago, we consume much more energy per capita because we have built a technological society around countless machines which dissipate energy. The more technological we have become, the more energy we have needed. The greater the dissipation of energy, the more likely we are to run into supply problems. To say either that the world needs a growing energy supply or that the society is becoming ever more technological is close to saying the same thing. We can solve the energy problem by discovering new energy sources or by conserving energy, using it more innovatively to accomplish needed tasks in homes, offices, industry, and transportation. The energy problem has its economic repercussions, its political and social aspects, and its international dimensions. But to achieve an adequate energy supply, we must view the scientific and technological aspects of energy as substantive, not side, issues.

When new science is applied to create additional energy, the social-political interactions are not necessarily diminished. For example, developing nuclear technology to generate electrical energy has given rise to new and exceedingly troublesome government-industry-public relations. So necessary and difficult is safe control of the dangerous materials through their entire life cycle from uranium mining to disposal and reprocessing of spent fuel, so drastic can be the consequences of carelessness in the design and operation of nuclear reactors, that all governments have had great difficulty organizing to encompass these burdens, all previously nonexistent.

To ensure adequate energy requires further development of the technology of energy sources and uses. Of course, we could deliberately choose to emphasize nontechnological approaches. To conserve energy, Americans could all wear long woolen underwear in the winter. Also, in theory, we could shift to mass transportation instead of relying so much on individual autos in our cities, a plan apparently no easier to pull off than the long underwear is to pull on. Woolies could be made available without a technological breakthrough. But could we in any way avoid the technological route to the attainment of adequate urban transportation? Say we were to choose, as an extreme nontechnological alternative, to go back to horses. We would find quickly that the horse-based system of 100 years ago could not be extrapolated to accommodate our needs today without major technological advance. To create the required 200 million horses would mandate scientific and technological developments for their mass breeding and feeding. We would also need machines for the removal from the streets of the enormous emissions, a technological problem tougher than designing afterburners on auto exhausts. With our present urban population and societal patterns, technology must be employed to move us about, whether by private cars, mass public transit, or animals.

What is true of energy and urban transport is also true for numerous impending natural resource shortages. Our technological society has led us to use up minerals once conveniently located near the earth's surface in America. The technological society requires steel, but we cannot make steel and essential steel alloys without chromium, cobalt, manganese, molybdenum, tungsten, and vanadium as well as iron, and we are becoming dependent on other nations where these metals are still accessible.

In a peaceful, cooperative world the safeguarding of our supply of minerals could be based on viewing the resources of the entire earth as available at market prices. However, in times of emergency, like war or the emerging of an unfriendly cartel, we might find ourselves dependent on output from our American mines or our stockpiles.[2]

---

[2] When Zaire was invaded by Cuban troops in 1978, cobalt mining there came to a halt. In a few months the world market price for cobalt had risen by a factor of seven times.

Most of the land in American not already being mined, or mined out, is owned by the government. We want to protect public lands from overexploitation. Also, we want to be certain that cost-effective exploration is carried on and adequate inducements exist to maintain a healthy minerals industry in the United States. Finally, we would prefer not to have to subsidize that industry but rather to allow world market prices to work to the economic advantage of the nation as a whole, including ensuring adequate stockpiles and industry production in an emergency. This is not easy to arrange because it takes years to develop an exploration, mining, smelting, and finishing industry adequately backed up by scientific and technological advances, and this cannot be created on demand when an emergency occurs or impends.

We could shift to mining less rich or more deeply located ores here in home territory, but they would be harder and more expensive to discover, remove, and process. Anyway, to accomplish any of these steps, the underlying science and technology would have to be pushed forward. In time, we might expect to invent resource substitutes, learn to recycle more efficiently, and design new ways to produce what we need with less dependence on specific minerals in diminishing supply. Once more, this would require more research and development. Scientific and technological advances have given rise to the technological society, which in turn creates shortages of resources and then puts on us a requirement to generate more scientific discoveries and technological advances to solve the shortage problem. The free market can act to effect solutions in part, but only the federal government can do the final balancing job and set national policies in the resource area.

Let us shift now to our second example—safeguarding air, food, water, and land—which has emerged in recent decades as a priority issue for our society. Humans have always contaminated their environment, but what is singular about our present situation is that we have reached the point where the incremental negatives of further technological implementations threaten to become greater than the incremental positives. We know this now and are unwilling to accept a careless impairment of health from contaminants in what we breathe, drink, and eat. Again science and technology are central factors. For the first time in humankind's existence, certain areas of science and technology must be vigorously advanced precisely because to do so is the only way to curb hazards that themselves are the results of our technological activities.

Thus, to meet the needs of an expanding population, we must achieve steadily increasing food production. Technological solutions—fertilizers, pesticides, preservatives, additives, and synthetic substances to increase animal protein output—can make possible this increase. But they may generate bad corollaries, such as a higher cancer rate.

Achieving safety, health, and environmental protection in today's technological society must be seen in an entirely different light from what illuminated this subject 100 or even 25 years ago. It means organizing to investigate, judge, and police. It means involving the government heavily in balancing the good that we intend against the bad that may result. To minimize disbenefits requires more understanding of the basic scientific phenomena involved. Only with more scientific effort can we have our food and eat it too.

Our third issue, health care, has given birth in recent years to a whole department store of dilemmas, with science and technology thoroughly entangled in them. Taking care of the ill is now compelling unprecedented public attention because a large discrepancy has built up between our expectations of medical services and the nation's realistic ability to provide them. An unsettled role for government in this area plagues us. Speedy communications to the public of advances in medicine have produced an insatiable demand for instant availability of every new aid. At the same time, we also have become a highly litigious society. Malpractice suits are now so common that physicians and hospitals feel forced to bring each publicly disclosed innovation for diagnosis or treatment into play early. Thus, when a breakthrough in medical technology occurred in which x-rays and computers were combined to provide three-dimensional scanning of the interior of the human body, every hospital immediately had to buy one of these expensive installations.

While medical services have been improved by science and technology, with the potential of even greater quality of health services ahead, the costs of medical care have soared. Management advances have not paralleled technological innovations, and better decision-making mechanisms in establishing priorities have not been introduced. It is often said that science and technology have priced medical services out of the market and that hospitals are too expensive for the patients that fill them.

Under such conditions, two extreme approaches are often proposed. One is to remove the government from the operation entirely and let the free market prevail. Individuals would buy only such medical services as they could afford, and the investment in technological apparatus by hospitals, clinics, and individual physicians would be determined by the paying market. The other extreme is to have the government take full control of all medical activities. A massive bureaucracy, surely soon dominated by short-term political pressures, would manage everything—from research at the biomedical frontiers to the operation of hospitals and allocation of physicians and equipment. In this extreme, unlike the previously mentioned one, the same medical services would theoretically be available to all. Since providing everything to everyone is not truly affordable, rationing would be needed. The rationing would become political, and the bureaucracy applying it would be filled with red tape, delays, and error. The quality of health care would be well below that intended, and the very sick often would wait in long lines.

But no matter how the nation tackles the problem, no program to arrange practical medical care for America can be implemented without science and technology as active catalysts and ingredients. Surely, further research on causes and cures of disease deserves to be an important part of America's health efforts in any case. Also, medical facilities will inevitably continue to be increasingly filled with technological devices. Cutting costs in the prevention and cure of disease will require more discoveries in science and further applications of technology. But in health care, as in other key issues before our society, the scientific and technological aspects are wrapped up with the social-economic-political portions of the overall problem.

To make matters even more difficult, some new health issues of unprecedented significance are emerging, spearheaded by advancing science. Researchers have cracked the genetic code, and genetic engineering now must be taken seriously. Rapid strides can be expected in learning the internal designs of living cells, controlling aging, producing new life forms, and finding cures to diseases where nature hitherto has been stubborn in giving up its secrets. To what degree must the government control experimentation in generating new types of life, such as dangerous viruses and bacteria, and in influencing existing ones, including humans, in the fertilization and gestation periods? If it is to be permissible to alter the human embryo, then toward what ends, to be decided by whom? Granted limited research resources, how much should we put behind the curbing of aging? Should we seek primarily to prolong the lives of diseased persons, or should we concentrate effort instead on understanding why illnesses occur? Again, it is not sensible to try to separate the scientific aspects of these issues from the social-economic-political ones for isolated analysis and decision.

Technology of, by, and for the People

At the beginning of this chapter, we asked why it is so confounded difficult to arrange that we make the best use of science and technology. The foregoing examples illustrate that we cannot isolate the scientific and technological components. The aspects are enmeshed with the important social issues and serve both as problem sources and as potential cures. Further, we see in each of these examples, and in many others if we had described them—urban decay, foreign trade, crime, education—the imperative need for organizational innovation.

If we want science and technology to be employed to benefit our democratic nation, why not organize so that the citizens make the decisions on the application of these tools? Shouldn't the public know best what it prefers? In America we come closer than does any other group in the world to possessing a system for enabling technological advance to respond to the will of the people. We are partly a free-market society and partly a government-controlled one. Opportunity is available for those with investment capital to risk it in developing and offering technological products to our marketplace. Consumers are privileged to purchase or reject these products, thus determining which technological approaches will survive while also creating incentives for other investors to satisfy market demands. The government sponsors other technological implementations, and government is directed by individuals we choose at the ballot box. We call the shots on what the government does by electing those we think will carry out our desires.

In theory, a truly free marketplace operating in parallel with decision making by a democratically elected government should lead to the technological advances we want. In practice, the market is not always free and actually is severely limited in important areas of technology. Government policies are often so loosely related to the democratic process that a citizen's influence is little and late. In matters technological, the structures of free-market operations and government activities are heavily intertwined so that neither the free market nor the ballot puts the citizen in full charge. It is true that if the American public comes to disapprove of any trend, it will eventually veto it. However, many years may elapse before this happens, and we may move far in ultimately unacceptable directions for long periods.

But none of this challenges what ought to be our objective. The design of our technological society should benefit humanity. The citizens should determine the objectives, and the system should operate to satisfy them. If it is not doing so, we need to innovate to make it work better. To improve the system, we first need to understand it. Step one in this understanding is to accept that America is a hybrid society, part free enterprise and part government-controlled. It has always been so, and the case can readily be made that it will continue this way for the forseeable future. Americans are not going to change to the point where only a few will be interested in the freedom to work and live as they themselves choose. They will neither lose their independent and competitive spirit entirely nor accept total dictation of their lives by the government. Thus free enterprise will long remain with us, whatever its momentary strength and influence on the nation's affairs. Similarly, it is naive or fanciful to believe the government of the United States will cease to affect technological advance powerfully. We shall remain a hybrid society.

To make this society work, the right roles must be assigned to free enterprise and to government control. But no easy, single rule will tell us the right roles and missions for the private sector as against the government's responsibilities. The nation's affairs, technological and otherwise, are too many and too diverse for one universal formula to set these boundaries and functions. The relationship between government and the private sector will alter with time and should vary from one phase of the nation's activities to another. That there is room for improved understanding of all this by the public is evidenced by the vehemence and the number of those whose proposals for solving all the nation's problems are wrapped in simplistic zeal for either total private enterprise or all-out government control.

Sometimes it seems half the nation believes the private sector equates with big business, which cannot be trusted with anything important to society because, in its selfish pursuit of "unconscionable" (more recently the favorite vitriolic adjective is "obscene") profits, it is ready to consider any actions, even those that impair society. These citizens think free enterprise perhaps may be a fine way to realize mass production, but private business should not decree what should be produced. If anything serious has to be decided and managed in America, then our elected government should take on the task. Those who think this way demand of candidates for office that they say how the government will solve all the country's problems once they are elected.

A second half of the populace, it often appears, looks upon the government as a totally inefficient, inept, incompetent bureaucracy beset by waste and graft, a segment of our society that handicaps our operation and contributes nothing positive. Furthermore, government is seen to lack the expertise to deal with science and technology. As this second half perceives it, if the government could be put out of the act and if everything then were left to a totally free market, all our difficulties would vanish. Private enterprise, this group determinedly hammers at us, is what made America great.

(A third half of the population, since more than one vote is allowed in this somewhat exaggerated but useful imaginary poll, lacks confidence in both the free-enterprise system and the government. With a deeply negative frame of mind, these folks expect no societal benefits from the activities of either.)

Granted a hybrid society, we need to arrange the proper missions for the government and for the private sector. Tackling this organizational task starts with recognizing that while there is a mission for government in almost every area, it is rarely an all-encompassing assignment that leaves nothing for the private sector. However, once we allow that government must regulate our technological society to some extent, we immediately must call out an additional major issue, one involving science and technology heavily, an issue we introduced in Chapter I when we spoke of the triangle of society-technology-liberty. This is that government regulation inherently equates with some loss of liberty for the citizenry. It can range from a clearly acceptable loss (such as the freedom to burn down a neighbor's house) to such severe interference with free enterprise's potential contribution to society that incentives are lost, innovation is discouraged, and what happens is not responsive to the people's real desires.

We must also recognize a surprising future danger regarding government control of the nation. As technology progresses, it will be more, not less, practical for the government to direct the country's activities in ever greater depth. This is because ever more versatile and powerful electronic computers and ubiquitous communication systems will rapidly increase the government's ability to acquire and process the information that keeps the nation moving. Assuming a regulation-prone government comes to be in charge, the new information technology will make it easier and more tempting to centralize the control of the information flow and disseminate precise and detailed directions to all as to where to work and live, what to produce, and so on. We shall go into the crucial relationship between advancing information technology and our society's organization in a later chapter.

Limitations of Free Enterprise

Free enterprise and technology teamed up long ago to produce an enormous flow of technological products. Capital at risk, seeking a maximum return, has employed the marketplace to match the opportunities created by technological advance with the desires and needs of the people. The whole is still a sound and useful idea. But the free-market concept alone will not provide for the fullest employment of science and technology on behalf of humanity. For some needs that technology can fill, and some attractive potentials society can reap from technology, free enterprise will not nurture the seed and at times will not even plant it. Furthermore, left to itself, free enterprise is unsuited to minimizing the detriments that can accompany technological implementations.

If free enterprise is based on investment at risk, the incentive for the investment being the anticipated return, then the ratio of the risk to the return is a succinct way of measuring the merit of a free-enterprise approach to an opportunity. When the return appears attractive compared with the risk, the free-enterprise approach will work well. If, on the other hand, the risk seems disproportionately high compared with the gain that might be expected from taking it, free enterprise will militate against resources being made available. The system will cause capital to be placed instead on other bets where the odds are perceived as more favorable. This concept of a risk-return ratio is a reliable guide to understanding the relation of free enterprise to technological advance, leading us to see why free enterprise has advantages in spurring technological advance in some areas and practical limitations in others.

For example, certain technological projects involve an investment—in the billions of dollars—beyond the means of the companies having the expertise to deal with the technology. Thus thermonuclear fusion (the basic principle behind the H-bomb) offers the possibility of some day providing virtually limitless electric power generation using essentially free and inexhaustible fuel (isotypes of hydrogen found in ordinary sea water) with minimum dangers and costs. This revolutionary idea may work out in practice, but again it may not. Despite enormous and fully competent effort, some fundamental scientific obstacles or engineering difficulties or economic surprises may show up that will block the successful completion of the development. Finding out for sure would appear now to require decades (following the two already used up) and may cost several tens of billions of dollars, substantially more than the net worth of any corporation that from other standpoints would be interested in this very specialized technological concept. Of course, in theory, a number of large companies might join forces to share the risk. However, for the potential return on investment to be sensible, the syndicate would have to own the resulting technological know-how and, rating it as highly valuable, maintain it under wraps as proprietary art, essentially planning for a monopoly in the field. This would violate the antitrust laws. Not surprisingly, controlled-fusion R&D so far has been government-funded.

In projects combining high risk and huge investments, the period of time required for the project to mature is quite often too long as well. Revenues may not commence until well beyond the first decade or two that comprise the research and development phase. The period when initial losses turn to earnings, if ever, may occur later still. Moreover, a project that large is almost certain to hinge for its success partly on government policy, notoriously fickle. Even if the governmentally influenced environment for the program appears favorable, both economically and politically, at its inception, those taking the risk will know it is impossible to predict what government policy will be in ten or twenty years.

The nation's synthetic fuels program is an excellent example of this. Four administrations—Nixon's, Ford's, Carter's, and now Reagan's—have felt it to be in the long-term interest of the nation that an industry be created to provide liquid fuels from shale and coal, of which the United States has centuries of supply, as alternatives to petroleum. Private corporations, it has been felt, lack adequate return incentives to move fast enough and are not in a financial position to accept the very high investment risks involved. Accordingly, a synthetic fuels corporation was set up by Congress and authorized to commit over $20 billion in goverment-backed credit as a first step either to guarantee loans for projects or to subsidize pricing for the synfuel products.

In response, a number of private enterprise teams were formed and began committing themselves to large investments with diverse degrees of planned reliance on the announced government aid. By the time a few years passed, however, most of those private companies had bowed out. Even Exxon, the nation's largest corporation with revenues and total capital over the $100-billion mark, decided to abandon its shale project after expending billions. Apparently, with all of the unknowns about how petroleum prices will move in the future, the extent of technological development required for success, the nature of future government-imposed environmental regulations, the international demand for liquid fuels, and numerous aspects of government policies involved in bringing such projects to maturity, the ratio of risk to return on investment was simply seen as too great, and the time to reach a potential profit position much too long.

If it is important to the nation that there be a synthetic fuel industry, it is clear the government will have to remain in the picture as a backer. Those few programs in the private sector that have not been terminated are counting on continued government help.

Projects to recover minerals from the bottom of the oceans also fall in this category. Here the United Nations has adopted a treaty governing ocean mining, which treats the nodules of copper, cobalt, nickel, and manganese on the ocean floor as "the common heritage of mankind" and creates a global mining authority to issue licenses to mine the sea. Although the United States has voted against the treaty, American mining companies probably will decide it is prudent to assume that America in the end will essentially abide by the treaty's terms. Given the very high cost of ocean mining—it has not yet ever been done on a commercial scale and the nodules are under 15,000 feet of water—it is likely to be attractive only if metal prices triple or quadruple. Even if a particular site is licensed to a private firm and turns out to be profitable, then, according to the UN plan, the new seabed authority will take up ownership of the adjacent regions, thus limiting the ability of the original pioneering company to utilize its expensively developed technology over the broadest possible region.

If it is important to the United States that dependence on imports of these critical metals be reduced by adding deep ocean sources, then the government will have to subsidize ocean mining in whole or in part. The ratio of risk to return does not make the approach a favorable one for private investment alone.

Another example may be found in the scanning of the earth's land resources from space satellites seeking data advantageous in the acquisition of rights to the resources pinpointed. Here part of the problem is that the technology is unproven, and more time and money are required to demonstrate fully that there is indeed commercial value to information about the earth obtainable from space. It is also not clear how a private corporation, even after making the investment and obtaining useful data, can market the information profitably to those who might find it valuable. This prospect is discussed more fully in a later chapter dealing with space. It is sufficient to note at this point that if the potential benefits of this application of space technology are to be realized and if free enterprise is to play a principal role, it will be only after the government has spent billions of dollars in the field.

Almost any new commercial airliner project, still another example, requires a multibillion dollar investment and a long time to realize financial returns. So high is the risk-to-return ratio, considering the size of investment required, that almost no American airplane manufacturing corporation has ever been equal to the challenge without leaning on related government-funded military airplane contracts. To illustrate, imagine that an American supersonic transport, with performance superior to the existing British-French Concorde, is believed to be an economically sound project. If such a program were started from scratch—with no government subsidies and no coverage of any of the costs through similar military airplane projects—the risk investment required before a profit period could be reached might easily grow to several billion dollars. This level of funding would be required to cover research and development, producing and testing of enough proto-types to thoroughly prove out the design, the setting up of production facilities, and the paying of manufacturing costs for a substantial initial production run prior to the receipt of funds from committed customers, who would pay only after delivery.

The limitations of a free-enterprise system operating totally alone are most evident in military weapons systems, where very little private risk investment, compared with total revenues, has ever been made. To make the point, let us fantasize that in the early 1950s some bold entrepreneurs with billions of investment capital behind them assessed the opportunities in intercontinental ballistic missiles (ICBMs). These risk takers, let us say, looked at anticipated advances in nuclear bombs, guidance electronics, rocket engines, missile structures, and the rest of the complex technology involved. They predicted that an attainable ICBM would win out in ten years as the basic strategic nuclear deterrent system for the United States. Seeking to put themselves in a position to reap an excellent return on their investment when that day came, they planned to sink the required billions of dollars into an R&D program. In addition, let us imagine, they thought to acquire property on the Florida coast and make arrangements with the nations who own islands in the adjacent Atlantic so as to be able to implement a suitable test flight program. They also recognized the need to invest in production and testing facilities adequate to construct and prove the numerous quantities of intricate hardware components making up an operational ICBM system. All in all, they expected to complete the development in ten years, including ample flights to demonstrate all performance aspects of their missile system, such as range, accuracy, payload-carrying capability, and reliability.

At that point, their plan was to turn to the U.S. Department of Defense, the single customer for whom their product would have been intended, and offer to supply ICBMs over the following decades at a price yielding a good return on the huge investment they would have made, around $10 to $20 billion. The American military, so they assumed, would see that what the entrepreneurs would be ready to deliver was precisely what the nation would then need. The Department of Defense would gladly sign a contract, and the enterprising company and the DOD would both be happy.

Any management group that proposed this multibillion-dollar, decades-long risk scenario to its board of directors would have been retired within forty-eight hours after the first presentation.

The government is the sole customer for military technology. It controls the market and the policies that create it. It determines the available budget and sets the priorities. The government role is indispensable. The field of military technology is not one in which free enterprise can meet the nation's needs. But the private sector's role is also indispensable.

Military technological systems and equipment constitute a large part of the total technological operations of the nation. The repercussions of this heavy participation of government go beyond national military preparedness, because technological developments for civilian purposes and international competition in civilian technological products are tied substantially to the trends and status of military technology. Thus the entire structure of America's technological industry is greatly affected by DOD's technology procurement and R&D sponsorship. The assignment of missions between the government and the private sector in military technology is an extremely important aspect of the strength of the U.S. economy as well as the country's security. We shall find it necessary to investigate this subject more in a later chapter.

If the government creates the market—this applies not only to military procurement but also to a good fraction of all the nation's activities in nuclear, space, oceanography, energy, weather, transportation, and health care technology—the risk-to-return ratio may be too high to attract private investment. The risk is heightened by the government's inefficiency and vagueness in defining its requirements and implementing its procurement plans. Short-range, highly political considerations dominate government policies, and these policies have been rapidly changeable and arbitrary. Members of Congress judge each and every project partly by the views and desires of their constituencies whose thinking goes to perceived self-interest. Big government budgets for defense or energy or space may be popular one year, unpopular the next. The funding is unsteady; it is decided on and made available at the last possible moment. Government is thus a poor maker of a market for its own purchases.

When government is the customer, sophisticated free-enterprise groups serving it demand arrangements that greatly limit their risks. In military technology efforts, this usually means complete government funding of all the research and development and of the initial production, the project phases where the greatest doubt attaches to the attainment of success. "Cost plus" contracting is common as is government furnishing of major portions of the needed specialized facilities for carrying out the contract.

In summary, when the government is the customer, which happens for at least a third of high-technology activities in America, only the ignorant would describe the process as a phenomenon of private enterprise.

Let us move now to consider a general limitation of free enterprise as the preferred route for advance of technology, namely, the weakness of long-range thinking in the private sector. Pension fund managers, bank trustees, security analysts, and investment professionals, who together control, directly or indirectly, most of the voting shares of corporations, have a near-term bias that is transferred to company managers, who then accentuate this tendency. When investors buy or recommend the stock of a corporation, they usually have in mind a time span for a hoped-for increase in market value of the shares from a few months to a year, sometimes two, but only rarely five years or more. This applies as well to the many individual small public shareholders of most corporations, whose actions are greatly influenced by the actions of the professional investors. It takes courage for anyone to buy shares in a company with the idea of waiting a decade to see a gain in market value, even if dividends seem secure. After all, the whole stock market picture may change drastically in that time, and so may the technology one is betting on. Steady dividends are usually in the 5 percent return range, so a rise in market price of a stock is important to the typical investor.

As explained earlier, this trend toward short-term goals is made much more severe by inflation. It is one thing to plan a long-term investment when the balance sheet is strong. But if the debt is increasing and the cash flow threatens to be too small to cover the inflation-swollen interest charges, extraordinary confidence must be displayed by the shareholders and the directors before a manager will take large, long-term risks, no matter how exciting the possibilities. The net result is that the free-enterprise system is better suited in practice to short-term advances in technology and exhibits important limitations as a vehicle for long-term, risky technological projects.

Some Unnecessary Roles for Government

We have just recited some limitations of free enterprise for implementing useful technology. When particular advances of technology are seen as in the public interest but are not well-suited to free enterprise acting alone, it may be expected that the government will become a sponsor or partner. Some find it natural to suggest that the government should step in not only in instances where free enterprise seems inappropriate but also in cases where free enterprise exhibits shortcomings. For example, it has been pointed out that U.S. industry has been lagging in productivity increases for well over a decade. Moreover, in certain important product areas, such as consumer electronics and machinery of many kinds, producers in other nations have been taking the world market and even our domestic business away from American manufacturers. Some conclude from this that our industry now lacks interest and competence in innovation and that we Americans as a whole have lost our Yankee ingenuity and our motivational incentives. The slip from technological leadership by the United States is perceived as a catastrophe deserving emergency, high-priority action.

Believing the government must take action to solve all problems, some have dreamed up expensive government programs and proposed new bureaucratic agencies to spur technological advance in the private sector. As an example, it has been urged that the government provide special tax credits to cover the cost of R&D in industry whenever such R&D leads to a "breakthrough."[3] Presumably, it would work this way: Industrial firms would come to a new government unit and there submit descriptions of completed R&D projects for consideration for favorable tax treatment of the funds spent. Each such candidate program would then be assessed by the government to see whether it qualified as a real breakthrough rather than merely a minor refinement. Companies with winners would receive government funds through the award of tax refunds and thus would be motivated to invest in more innovative effort.

The necessary judging of the proffered projects by the government would be an interesting challenge in itself, involving undefinable criteria and arbitrary, subjective evaluations. If taken seriously, this proposal would require that industry employ engineers especially to interface with the government, and the government would have to assemble a comparable group so that the two groups could negotiate. The expenses of all this paper and discussion effort would be borne eventually by the consumer, using up funds that might have been devoted to real invention. Moreover, the costs would decrease the net return on investment and discourage such investment. Whenever a new technology program is created with government funds, the money somehow has to come out of the private sector, subtracting from the capital otherwise available there for innovation. Moreover, subsidies through tax credits conferred by the government put the government in the position of selecting the areas to be pushed, and this militates against advances in other areas selected by the free market.

Eager to spur creative technological advance in America and noting that small technological companies seem to originate a disproportionate fraction of innovative products compared with large and long-established corporations, some press the argument that the government should subsidize the founding of more new small companies. But these folks have the cause and effect inverted. It is the new technological ideas that give rise to the new small

---

[3] Through its programs in defense, space, energy, medicine, etc., the federal government already is the sponsor of half of all the nation's research and development.

companies, not the other way around. Individuals—often young inventors who propose radical departures from present technologies, or mature engineers and scientists who depart from older companies to pursue their new ideas—launch the new ventures.

Successful small technological corporations naturally have a higher growth rate and create more new jobs per capita than do large mature corporations. All the jobs in the former group are new, whereas jobs in the latter group are the sum of new and old ones. Today's large technological companies were originally much smaller and were based on technological advances occurring decades before these companies attained their present size. Today these large companies, even though they include much R&D and are producing new products and new jobs regularly, have a total sales volume and employment level dependent mainly on older and, in some instances, declining products. They may create far more new jobs than the smaller new companies, but at the same time they may close out more old jobs. Thus they may exhibit a declining overall employment rate while still making the major contribution to the nation's incremental technological employment.

Despite inflation, high interest rates, recession, and a low or uncertain stock market, a new technological corporation can get started in the United States with private funding if it has a good enough idea and a strong team.[4] Those who have difficulty financing the start of a company do not possess the combination of technological innovation and engineering and management competence to rate high enough with potential investors, who choose other investments instead. If the government should enter as a financial backer, private sources will support the attractive candidates as now, and the government will back the ones deemed less likely to succeed in a distortion of what otherwise would be a superior free-market rating system.

In seeking the soundest new applications of science and technology, we must recognize that R&D is only the tip of the iceberg. Other factors, such as matching the technology to the market and arranging for efficient production and distribution, usually require much more management breadth and investment of capital than the initial R&D. The private sector is more suitable for judging and handling these aspects than the government, which is at best an incompetent third party standing between the private sector and the ultimate customer. For most goods (not ICBMs or nuclear bombs), that customer is best represented by the free market, not by a government

---

[4] The United States has some 300 venture capital firms. It has been estimated that they examined around 10,000 proposals from entrepreneurs in 1981 and backed over 300 technological companies that same year.

contracting officer. The market will make the best decision as to what new technological applications should be developed.

There is no evidence that Yankee ingenuity is disappearing. Nothing has happened to lessen America's fundamental talent for innovation. The best thing the government can do to spur technological innovation in the United States is to improve the economic environment; cut inflation and cure recession; stop excessive spending, taxation, and regulation; and take better care of those functions which only the government can perform. Many such functions exist. Let us cite a number of them.

Some Necessary Roles for Government

One important and mandatory mission for government is the sponsoring of basic research in the universities. Fundamental exploration into the laws of nature, carried on without a parallel motivation to exploit the results, is never the main goal of a private company, which seeks instead to apply science and technology to yield a good and preferably early return on investment. To be sure, intriguing scientific questions may often arise during industrial product development which, if pursued determinedly to find the answers, might qualify the activity as pure research. Scientific discoveries might then be made, some even deserving Nobel awards. However, when an industrial concern allows its researchers to engage in such basic research efforts, rather than in R&D focused on its product endeavors, it is most often as a tiny fringe effort. Usually it is best labeled a philanthropic contribution to society.

Some pure research conceivably might be funded deliberately by a company to further the standing of its R&D staff in the world's elite scientific research fraternity and to obtain ancillary benefits of that enhanced status over the long term. All corporations, even those that are very large and in a specialized field of endeavor (computers, chemistry, metallurgy, photography), are aware that the total scientific and technological advance going on in the entire world dwarfs their own, and they arrange for a steady drawing in of ideas and information generated outside the corporation. Communication paths to other researchers must be constantly traveled and new ones opened. However, it is a rare American corporation that engages in truly basic research.

Such research investigations are ideally suited, on the other hand, to the objectives of university science and engineering departments. It is highly advantageous for students, particularly those seeking higher degrees, to spend their university years in an atmosphere of research into the fundamentals underlying the profession they are preparing to join. Also, when engaged in frontier research in their specialties as well as in lecturing, professors become more expert in the basics they teach.

When America's technological corporations provide funding to university science and engineering departments, it is naturally to sponsor efforts in areas of their interests. However, in pursuing basic research, the university researcher is motivated to make discoveries, be creative, solve mysteries, understand fundamentals, all quite independently of eventual applications, or even of interest in whether there might ever be any applications. If the corporation's motivation is instead to advance the field so as to make a financial gain, then the projects of necessity will be selected and limited to those in which such a return seems likely and clear. This translates to relatively short-term, incremental, nonspeculative advances, totally suitable to industry, but not properly called basic research. When industry, with this second kind of motivation, sponsors projects in universities to be carried on by individuals with the first-named motivation, the match is poor indeed.

If all university research were financed by private industry, industry leaders would choose the subject matter for the investigations and the largest, richest companies would have the biggest say. They would judge the researchers and their outputs by their own criteria, continuing some programs, dropping others. The efforts of the nation's corps of scientists engaged in advancing our knowledge of nature's secrets would then be set by industry managers committed to advancing the return on their investments in their specific, narrow fields of endeavor. Some programs in the nation's overall best interest would nevertheless occur, but inadvertently and not in deliberate response to national objectives.

The government, and not competitive industry, is the proper and natural source for funding university basic research. Because it benefits all citizens in the end, it is right for all citizens to share the costs. There is also another very practical reason why government should consider itself the permanent chief source of sponsorship for university research. That is because strong university research is necessary for strong higher education. It requires no deep analysis to appreciate that education in science and engineering is a fundamental pillar supporting a technological nation. Admittedly, it is not guaranteed that science and technology will be well applied in the interests of our country because a stream of Ph.D.s leave American universities each year equipped with basic skills in mathematics, physics, chemistry, biology, and engineering, not even if there is a parallel outflow of humanists and a drop in the production of attorneys and accountants. But things will go badly for America in the future if, as now, we continue to graduate fewer engineers annually than Japan, whose population is less than half ours.

In theory, with a healthy free-market economy, systems of higher education would be responsive to the demand for graduates. With our technological industry pulling away many of the best faculty members and some of the lesser ones as well, offering much higher salaries in an attempt to meet its rapidly expanding short-term staff requirements, a future shortage of engineers is bound to result. However, because engineering salaries then will rise, the growing shortages will enhance greatly the opportunities for the new graduates of that time. The high opening salaries will become known to entering freshmen choosing university majors, and the demand for engineering education will go up. Families will be more willing to go into debt to finance their sons and daughters, who will take part-time jobs as well, stretching out their educational period as required, knowing it is all worthwhile because when they get their science and engineering degrees, their careers will be satisfying. Student loans will be available because the prospects of repayment will become excellent. After some years, the gap thus will be closed. Some overshooting may even occur next, with too large a fraction of all students deciding on an engineering or science career.

The resulting ups and downs of availability of engineering and science graduates is not the most serious defect of the scenario just pictured. It is rather that the process takes too long and the nation's position will suffer greatly in the interim.

Today federal and state governments are reluctant backers of higher education in science and engineering, and budgets are being cut. This is not sensible. When government spending must be reduced, a simple priority rule should apply: Cut that which is expensive and contributes little in the long run; retain and expand that which is inexpensive and contributes greatly over time. Education in science and engineering is in the second category, yet short-term thinking is causing the government to violate the simple priority rule.

The technological industry should be a strong backer of higher education, but government alone represents what should be the long-term priority interests of all the citizens. Some business leaders are completely convinced that higher education is a permanent requirement for the nation and that it is a good idea for industry to put aside part of its budget for philanthropy to help ensure that the process of higher education is steady and of high quality. But industry as a whole has trouble handling conflicts between its short- and long-term requirements, between very direct investments exclusively beneficial to a specific corporation and indirect ones that help society as a whole, including the competition. Industry's contribution to funding higher education thus has been (and even with substantial expansion in such aid now being advocated by some industrial leaders, surely will continue to be) well below the financing needed. Government is the logical source of continued financing.

Let us shift now to another necessary government role. We mentioned earlier that one mandatory function for government is to participate in controlling the potential negatives of technological implementations. But satisfactory government regulation of the negatives of technological activities is more easily described than arranged. Underregulation and overregulation are common, and this not only handicaps the nation's economy but creates a bottleneck in the beneficial use of science and technology.

From the invention of tools and the first use of fire for cooking to the operation of nuclear reactors, every technological activity implemented to attain gains has had the potential of producing detriments as well. Zero risk does not exist. Whether to allow certain drugs, pesticides, vehicles, building materials, paints, etc., on the market comes down to value judgments. For some things the best way for these judgments to be made is through the free market, without the government's involvement. When persons select an automobile, buy cigarettes, take a trip on an airplane, or purchase an over-the-counter drug or a bottle of whiskey, they have some awareness that they are taking on hazards along with the benefits they seek. Items that offer customers what in their personal judgment is an acceptable balance of the good and the bad will be purchased. Those the public views as offering too great a risk for the envisaged reward will lose out in the marketplace. But for numerous areas the free-market approach by itself is inadequate. A trial-and-error process, in which the consumers' past encounters with positives and negatives guide their future purchases and control the risk-taking balance, may take too long, with dire consequences during the experience-gathering phase.

Even if the free-enterprise sector were led totally by management teams of the highest integrity and public spirit, why should those managers presume to set the compromise between economic costs and dangers? A drug manufacturer should not decide for the nation whether a new drug is ready for the market. An individual automotive company could hardly limit itself to producing a car that is as safe as it knows how to design, because its selling price would be prohibitive. A tobacco company should not be expected to liquidate its business voluntarily because of the established health hazard of its product when the law of the land does not require this and tens of millions of Americans are willing to accept the risk to their health in order to satisfy their desire to smoke.

If the free-enterprise components of our society had the entire responsibility for balancing the risks to safety, health, and the environment, the conflicts of interest would be unacceptable. The people expect the government to integrate the diverse value judgments of the citizenry and to provide a degree of objectivity which they correctly believe the private sector's self-interested suppliers are not capable of. If there were no government regulation of hazards, then responsible companies would be handicapped in competing against irresponsible ones. The latter would be able to offer their less safe products at lower prices. While customers would stop buying once the defects were discovered, and suits for damages would help drive the unscrupulous competitors out of business, still, with no regulatory guidelines and policed standards, new unethical entrepreneurs would pop up continually. They would take again the same short-term gambles, contributing upsetting forces to the market. The enlightened leadership of the free-enterprise segment of our hybrid economy recognizes the need for government regulation as much as does the public.

Government regulation is necessary and, properly done, can help protect safety, health, and the environment. Government regulation is necessary in certain other technological areas as well. A very critical area has to do with reaping the bountiful benefits of information technology. Information is what makes the activities of the world go around. Technological advance in the acquisition, processing, storage, and utilization of information can improve the way every human activity is pursued. Because of this, the computer-communications revolution inspired by information technology will be the most powerful technological advance with which society will have to deal over the next twenty years. If we do not destroy ourselves through the irrational use of nuclear weapons, then it is the area of information technology that will have the greatest impact on shaping the next several decades.

But without refereeing by a government prepared to deal with the relationship of technology to society, the advantages of the computer-communications revolution will be gained only partially and slowly, and unnecessary disadvantages will develop. We shall demonstrate this in a later chapter.

Let's move to still another area. Government owns the resources for some technological activities important to the nation. Many of the mineral, oil, coal, and shale deposits of the United States are situated on state or federal land, and very often, so is the water supply needed to develop these energy sources. Lakes, rivers, seaports, and some railroads—all essential to industrial transport—also are government-owned. Moreover, full development of government properties can give rise to severe pollution of the earth, air, and water. The federal government must set the environmental rules because what happens in resource development can affect the land and population far from the immediate vicinity of the activities. Allocation of water rights is even more difficult to arrange in some ways than mineral rights, because the government must continually balance uses of water for the purposes of energy and mineral development against applications in agricultue, chemical production, and urban use. Also, some complex international issues are present and permanently involve the government. For example, the demand for energy obtained from coal and shale depends greatly on relations between the United States and the OPEC nations.

Could we solve most of these problems by free enterprise, say, by having the government sell these resources to the highest bidders in the private sector as soon as potentially advantageous projects surface? Proper prices would be the discounted present value of the future realization of income from the properties; this value would be determined by the markets for the products, the capital investment needed for the development, and the safety, health, and environmental protection requirements. To be realistic about the most significant of these resources would be to admit that the sales price usually would be beyond the net worth of almost all companies. Similarly, the period of time required to develop the property purchased and obtain a positive cash flow to recover the huge investment usually would be much longer than would represent sensible planning for corporations that might otherwise be interested. Leasing of rights to the private sector appears to be the most practical approach.

Because government has a permanent role to play in the utilization of such government-owned resources does not mean it must dominate the development of such properties, leaving no opening for free enterprise. For example, the government does not really need to control and foot the bill for R&D in synthetic fuels. The private sector is capable, both technically and financially, of developing gaseous and liquid fuels from coal, shale, or biomass if three conditions obtain. First, the government must set clear, practical rules for safety, health, and environmental protection. Second, the government must be willing to grant leases for suitable government resources at reasonable rates. Third, a market has to exist for the products that will flow from the effort.

The government can contribute by strengthening the market, since it is itself a customer for fuel. If the government believes it is in the national interest to move faster on synthetic fuels than the free-enterprise system will do on its own, government can offer to purchase synthetic fuel. With the government willing to guarantee a realistic price for an agreed-upon quality and substantial quantity of output, free enterprise then will be willing to take the investment risk. Private industry has the expertise to judge the price-risk relationship. The lowest bidders in a government-held competition, everything else being equal, should win the privilege of supplying the government. This is certainly a more sensible role for government in synthetic fuels than out-and-out sponsorship of projects in which industry, with all its experts and its large cash flows, refuses to invest its own funds.

The energy field offers a good example of the need to set sound missions for both government and the private sector if technological advance is to benefit society most. It is also an excellent example of how wrongly chosen government missions can easily develop. It is especially true of energy, although far from being exclusively limited to it, that a crisis atmosphere has surrounded the subject, causing public confusion and leading to short-term, opportunistic political decision making that interferes with the workings of both the free market and the government.

The nuclear dimension of energy, though having singular characteristics of its own, is especially illustrative. Nuclear reactors to generate electric power bring to mind the atom bomb, dangerous radioactive materials needing tight control, classified military matters, highly advertised potential catastrophes, unprecedented potential liabilities for suppliers and operators, huge capital commitments, and long periods of time for development and before reaping returns on investment. All these items in turn bring to mind the government's involvement. It is unthinkable that we might have organized the nation's efforts on nuclear electric power in such a way as to have denied the government a hand in the game and have left it all to the free market. By the same token, it is not surprising that the government's record in management and policy formation can be readily criticized. Consider the case of the Clinch breeder reactor.

Years ago, when the nuclear reactor approach to meeting the nation's future electric power requirements was first launched, no OPEC petroleum pricing phenomena had yet appeared. The nuclear option seemed to make sense because it was deemed more economical than other alternatives and its potential for serious dangers was regarded as containable. However, looking ahead in time, it was feared a uranium shortage might develop. Readily available, low-priced uranium ore might run out, and we would have to turn to ore harder to mine and process. The proposed answer was the so-called breeder reactor, which could stretch out uranium supplies from decades to centuries. Tennessee's Clinch River project was created as a step in providing the nation with early breeder-reactor experience; it has cost about $1.5 billion so far.

However, much has changed since the inception of the project. The immediacy of need for breeder reactors has disappeared, the timing of the need is now seen as well into the next century, and large downward revisions have been made in estimates of growth in electric power demand. Nuclear reactor economics has also been revised, and the facilities are now seen as far more expensive than earlier estimates indicated, especially when added safety steps, now deemed essential, are included and the full costs of nuclear waste disposal are taken into account. An optimum breeder-reactor program today is quite different from the one that was started years ago. Unfrenzied, long-term research still seems sensible, but an expensive demonstration of an out-of-date approach does not. Unfortunately, it also happened that the managers of the Clinch River project underestimated both the time needed to reach various milestones and the funding required.[5]

Not surprisingly, then, the project has been judged widely and repeatedly in recent years as unnecessary, undesirable, premature, and wasteful. Yet it persists, suffering from a common ailment of government projects—that they may be hard to start in a timely fashion, but are impossible to stop once started.

Numerous alternatives to petroleum and nuclear-based energy have been put forth as deserving attention and government sponsorship in view of the expected eventual falloff in assured petroleum supplies under U.S. control and the controversy over the hazards and economics of nuclear electric power generation. These alternatives include conventional and unconventional utilization of coal, direct solar radiation conversion, ocean thermal (using the temperature difference between surface water and deep water) and geothermal (extracting the energy contained in hot gases and liquids deep in the earth) sources, windmills, and the turning of agricultural products into fuel. If the energy supply problem is a severe national emergency, then massive crash programs under government direction and

---

[5] The project, started in 1970, was to be completed in 1978 at an estimated total cost of $200 million. It now consumes $250 million per year and is estimated to be completed in 1990. David Stockman, head of the Office of Management and Budget in the White House, while a congressman, described the Clinch River breeder as "a technological turkey." More recently Congress's General Accounting Office estimated the total cost at completion to be more than twice the $3.6 billion estimate of the Department of Energy—or some forty times the estimate at the program's inception.

funding may be indicated to save time in providing energy alternatives. Land or water availability problems could arise concomitantly with these programs, and enormous pollution problems could accompany some of the approaches. These added considerations would also make direct government involvement mandatory. Unfortunately, if it appears certain the government will play a strong part, this adds to doubts and confusions about the stability of these programs and ensures that free enterprise, or entrepreneurial risk taking, will be at a minimum. Risk takers prefer not to gamble on what the government may do. This in turn almost guarantees that only by government sponsorship will the development of some energy alternatives take place.

Fortunately, the crisis atmosphere surrounding energy in the 1970s is disappearing. Except for basic research into the underlying fundamental physics, chemistry, biology, and engineering pertinent to energy phenomena (which should largely be performed at universities with government funds), free enterprise would appear capable now of providing the motivation, expertise, and investment necessary for moving most alternative energy developments at a sensible pace. Such an approach is better than a heavy dependence on government, since the government lacks expertise in the technology and is not skilled at judging the real market.

Thus, if the use of sunlight striking the roofs of homes to produce electric power is a sound alternative, the self-interest of homeowners and industrial producers of the equipment should be sufficient. Indeed, private investors are already backing the solar energy idea with annual funding in the several tens of millions, and the total of past private investment in solar energy is now up to several hundred million dollars. The government is the leading authority on the desirability of lowering our dependence on OPEC oil supplies, but is not as competent as the private sector in the technology and economics of photovoltaic cells and is not in a good position to gauge the attractiveness of installations to individual users.

Energy conservation steps taken in the United States since 1973 are estimated to have cut demand by 10 to 20 percent. Some serious estimates indicate that it is possible to increase the overall conservation of energy much more without handicapping our economic growth and that many tens of billions of dollars more in energy costs could be saved annually. Some feel that the government must be assigned the mission of financing energy conservation R&D and subsidizing installations to conserve energy. However, giving homeowners tax credits to induce them to install a solar-cell roof would appear to be no more justified than for the government to pay part of the cost of a new, fuel-saving automobile for those same homeowners. Similarly, since insulation additions in homes and buildings can often lead to savings in energy sufficient to more than pay for the costs of these steps, why should the government bribe homeowners to do what is already in their best interests, namely, to make investments that will pay off? If homeowners are reluctant to do what is financially sensible for them, won't advertising by insulation installers and loan companies educate them in a reasonable time?

Builders and homeowners generally have a poor understanding of energy conservation. The building industry is highly fragmented, and the design of buildings and homes, like the design of home appliances, caters to the customer's narrow focus on first cost alone and not on the more complex relationship of first cost to annual savings in power consumption bills. Consumers have not been found knowledgeable enough to buy somewhat more expensive light bulbs that produce the same light intensity with less power consumption and longer life. Thus the free market does not work rapidly in converting America's energy habits to energy conservation and cost saving. Eventually the facts will become known, and purchasers will choose the more economical route. But government can probably accelerate this process by appropriate propagandizing.

Government does not help by engaging in regulatory policies that underprice fuel and electricity. If additional consumption of energy, by government control policies, is priced on original (rather than on the inflated, true replacement) capital figures and on average (rather than incremental) fuel costs, then underinvesting in conservation is a certainty. The proper role for government is to stop doing the wrong things, rather than to do more things in all directions, most of which probably will also turn out wrong. In this case, the right approach is to allow market prices to determine costs to the consumer.

For a government bureaucracy to sponsor efforts to redesign industrial processes to save energy is especially ludicrous. What ought to be done depends on details of engineering and economics that industry understands best and that government should not be expected to understand at all. Government subsidizing of either housing or industry changeovers to conserve energy is not basically different from the government's decreeing the wearing of, and the government's paying for, warm sweaters for all citizens so that less fuel will be needed to heat homes and workplaces.

Let us shift now to a final example of a necessary government role. If our nation's technological activities were left entirely to a free-enterprise approach, harmful monopolies could be expected to arise. Keeping a watchful eye on potential or developing monopolies and breaking them up should they happen are proper roles for government. As society grows more technological, this requirement for government participation increases in importance. Simultaneously, it becomes more difficult for government to handle the monopoly problem.

If the government only had to be concerned with preventing harmful monopolies, it would have a difficult enough job in an increasingly technological society. But government also has to shoulder a different, seemingly opposite responsibility. It must sometimes allow and actually create monopolies if science and technology are to be used to the fullest. For instance, to compete successfully against industry teams of foreign governments in the world market, U.S. competitors may sometimes have to be allowed to team up. For certain technological activities (electric power distribution and telephone service are prime examples), it would be wasteful, impractical, even chaotic to countenance overlapping competitive enterprises operating in the same geographical areas. Picture only two independent telephone companies duplicating underground conduits and external wiring (and receivers on our desks) so that each company could connect with each of our homes and businesses—or a similar duplication of electric power lines. The only sensible solution is to grant monopolies to chosen entities in defined regions. When such exclusive privileges are bestowed by government, the activity must then be monitored to ensure that the rates charged are fair to the consumer and consistent with a reasonable return on investment so that capital can be attracted to extend and improve the system. The quality of service must be judged by the government as well, since rates cannot be assessed except in relation to the service rendered.

The necessary assignment by the government of monopoly privileges to provide technological services is tremendously broadened by the computer-communications revolution. This revolution is justifying the detailed cabling of our cities, industry, and government for the flow of electronic information that will ultimately affect every American's pattern of activity, from management, transportation, education, and professional activities to entertainment in the home. As we shall explore fully in a later chapter, this presents numerous new opportunities and problems, and the government is a required participant in handling them.

Government and Business—Adversaries or Partners?

We have seen that both government and the private sector are important in determining what happens on the scientific and technological front in the United States. The free-enterprise system's contributions consist of a myriad of individual increments. Still, the objective of each private entity is clear, namely, a favorable return on its investment and a cozy niche in the marketplace. By contrast, the government's duties are less easily definable. Even when the mission appears clear, it is not so easy to measure whether it is being accomplished well. The bottom line, net earnings, so useful in grading corporations, has no parallel in the government's technological operations. The report cards for government are often no more than the next election's results. By the time history takes the true measure of government performance, it is a bit late to alter the trends.

The government's roles in technology—such as setting spending-taxing policies to provide a healthy economic environment, sponsoring basic university research and higher education, fighting and granting monopolies, purchasing multibillion-dollar weapons systems, setting safety rules for nuclear reactors—involve a tremendous array of complicated interactions, with political crises overriding long-term objectivity. Benevolent, wise, and all-powerful dictators, with no elections to be concerned with, would still find it exceedingly difficult to organize effective government for a technological nation and to balance short-term considerations against long-term goals. This would be true, even if they called on their country's top brains to advise them.

Operating the technological society in America intelligently requires strong and defined roles for both the government and the private sector. But this does not mean that the best results will always be obtained if these two influential forces can be made cooperating partners. Sometimes they should be adversaries. A well-managed business entity's leadership will seek to meet its narrow and specialized objectives. These objectives may be partially in harmony with national goals but even then will be different in priorities and details. Sometimes industry aims will be opposite to what the government should or will be trying to do. Since neither the government nor any private entity will be perfectly led—wrong decisions will be made by each because of inadequate information and/or incompetence—the goals of the private sector and the government will often turn out to be at variance for this reason alone.

Since part of the government's function is to regulate and referee in order to protect the nation against improper private actions, government and private groups will often be natural opponents. The private sector itself has many separate elements, and they will not all have the same goals. Business entities often see things differently from one another, and a few even occasionally possess views greatly at variance with those of labor unions. The academic world meanwhile is usually off doing its own thing. And with short-range political considerations tending to dominate government actions, segments of the government itself will pull in many different directions.

Thus, to make copious use of the words "cooperation" and "partnership" in a general description of the preferred relationships between government and private enterprise is not realistic. Complete, coordinated harmony not only cannot exist in the real world, but in an ideal world it would not be the best way for national influences to interact. Granted this, however, America can be properly described as plagued with unnecessary and harmful adversarial relations between the government and the private sector. These relations constitute a major handicap to advantageous technological development.

We shall illustrate this and discuss measures to cure the problem in the chapters to follow. For now, it will suffice to state that the general cure for the spreading disease of deteriorating relations between government and business lies first in a thorough acceptance of the idea that the basis of the United States economy is a permanent hybrid of government control and free enterprise. Next, it must be understood that success requires arranging the best roles for the government and the private sector, despite the fact that no single, simple rule for dividing up the tasks can be invented and applied. Recognizing that the assignment of missions is difficult is at the heart of solving the problem. Depending upon the area—environment, monopolies, energy, computer communications, health care, transportation, inflation, education, basic research, economic competitiveness, national security—the assigned optimum roles for the two sectors should be expected to be different. The roles should be custom-tailored and altered as circumstances change and technology advances.

In the workings of the triangle of society-technology-liberty, technological advance constantly thrusts forth challenges to liberty and leads to modifications of societal patterns as well as to new opportunities. Free enterprise is a manifestation of the liberty focus of our civilization, but the society focus must deliver stimulation and constraints.

Chapter Four The Regulation-Innovation Quandary

Suppose we knew positively that no hazards would ever result from technological implementations. Pharmaceutical drugs would fight disease without side effects. Nuclear reactors would be inherently safe. Fingers could never get caught in better mousetraps or children's toys. Food preservatives and additives would keep foods fresh, make them tastier and more nutritious, and contain nothing harmful. What is more, while dreaming, let us say we could be confident this immunity from harm would always be. Technology would forever deliver only benefits, never disbenefits.

Even in that technological heaven, three criteria would need to be satisfied for sound application of technology. First, the technology would have to be ready or feasibly extendable. Second, since not everything that can be built, should be—even though the inventor is always convinced it should—the application must pass the economics test. The cost of the technological implementation must be exceeded by its value. Third, it must fit socially. The technology has to satisfy a real need or desire. Evidence that this social test has been passed is the favorable reception of the product by the free market, or its receiving an adequate priority by the body politic to meet a national requirement not filled by free enterprise.

The society just pictured, where technology bestows favors sans hazards, has never existed, although ignorance caused the world to act until recent decades as though all possible harms of technological advance were minor and tolerable. It is understandable that dangers in the workplace did not loom large when civilization had not yet reached the point of abolishing child labor, and malnutrition and numerous diseases now curable or preventable were more likely than were industrial accidents to destroy health. The prevailing culture then dictated that unemployed workers choose dangerous jobs over no job, and indeed the alternative of starvation would have been even more dangerous to their health.

In earlier phases of the technological society a product that might be harmful was vaguely thought to be self-eradicating, since it was assumed people would not go on buying what hurt them. Without need for governmental participation, physicians would discover negative side effects and properly weigh the potential benefits and risks for their patients before writing a prescription for a drug. The fledgling electric power industry knew it could not produce electrical power and distribute and sell it to businesses and homes if electricity were not safely containable, if exposure to it were not controlled through deliberate engineering design, and if people generating and using it were frequently electrocuted. It was thought that for all products, manufacturers and consumers together constituted an adequate and natural team to limit the danger.

Today's Need to Include Hazards

The situation today is different from many standpoints. The potential negatives of technological advance are now a factor in all three criteria cited: (1) readiness of the technology, (2) soundness of the economics, and (3) social acceptance.

The underlying science and technology must now deal with the dangers as well as provide the benefits. In the process of designing a new safety pin or a bullet train, a paper diaper or a contraceptive pill, a factory robot or a petrochemical facility, understanding, evaluating, and minimizing possible hazards must be seen as indispensable parts of the task. The overall science and technology of safety, health, and environmental protection are viewed as transcending individual technological projects and as requiring major effort and organization. It is accepted that national programs must answer such scientific questions as: What are the main contributors to air pollution? What regulations would curb the pollution to what extent? What basically causes various substances to be carcinogenic? Are we in danger from deterioration of the ozone layer surrounding the earth? If so, how can the danger be minimized? How serious is acid rain to crops and fishing? What harm can stem from the exhaust of vehicles burning diesel fuel? Any technological implementation now requires consideration of the science underlying its character in delivering both looked-for positives and unasked-for negatives.

The second criterion, the economics test, is now made more severe by the hazards issue. We still must ask whether the promised benefits are cost-effective, but now we must always include the cost of holding down the negatives, curbing the dangers that could result from producing the product and using it. A proposed technological development may pass the economics test as to benefits but fail that same test as to disbenefits. The product may lend itself to development and manufacture at a cost that makes its selling price economically attractive, but if it is too hazardous, limiting its dangers to an acceptable range may raise its total price to an unacceptable level. We may not even know initially how to decrease the disbenefits adequately, and the research required to learn how may not be economically justified.

A technological implementation that might have passed all three tests a decade or two ago may now flunk them all, including the third one of social acceptance. This last criterion comes down to a value judgement, one often difficult to anticipate and pin down, and one exerted by a larger segment of the citizenry than merely those directly involved in the product's manufacture or use. Thus the controversy about the safety of nuclear electric power generation brings in members of the public well beyond the workers within the facility or those who receive and use the reactor's generated electric power.

For some aspects of *safety, health*, and *environmental protection* (we shall henceforth often refer to these matters by the acronym SHEP), we seem to have reached saturation as to tolerable negatives. For instance, certain rivers cannot be allowed to be polluted further. (One in the midwest caught fire recently, and the flames picked up when the fire-fighting tug proceeded to spray them with river "water," which had become a flammable liquid fuel through heavy pollution.) The air in some cities, it is felt, cannot be permitted to carry more of certain harmful molecules. In such instances, if expansion is to be allowed in activities producing negative SHEP effects, then some existing sources in the area must first be eliminated.

Organizing to Attain Net Benefits

SHEP alternatives need to be compared. For instance, burning coal to produce electricity or converting it into liquids useful for transportation systems, will ease the energy supply problem. This is a benefit. However, the mining, processing, delivery, and utilization of coal bring SHEP disbenefits into play. The benefits minus the disbenefits, the net result we are interested in economically and socially, may work out to be a very small gain, one easily exceeded by going to other alternatives, such as a greater effort to conserve energy. Again, conservation offers the plus of energy saved but presents the minus of the expense to change over to new systems of factory production, building construction, or transportation that will use less energy.

Because of the growing appreciation that it will pay to net out the benefits less the disbenefits of all alternatives, economically and psychologically, this is becoming the new acceptance test for technological operations. The way we motivate technological advances and select particular ones to pursue should be made responsive to this idea of using net benefit assessment. Projects and products where the net benefits are small or negative should be held back, while projects with high net benefits should be pushed forward. To accomplish such assessments on a host of large and small projects coming forth daily for decision and implementation all over the country, we again have available the combination of free enterprise and government control.

Many decisions can best be left to the free market. Suppliers and consumers naturally will always exert a useful influence in seeking positive net benefits. The newly heightened sensitivity of the public to disbenefits, and the fact that simultaneously we have become a highly litigious society, will help the free market along in identifying and removing the high SHEP impairers. Today buyers are suspicious, conditioned to assume the worst, ready to complain or even sue if any product or activity harms them. At the same time, managers in industry are not surprised to find themselves in difficulty with employees or users if their products present hazards, especially those for which no clear warning has ever been issued. But free enterprise is insufficient by itself to meet fully the new positive net benefit guidelines.

Individual negative SHEP effects of products or operations may take too long to surface. In the interim, unacceptable damage may result (some of that damage being to the net worth of the producers held responsible in later court procedures). Sometimes the harmful contribution of a specific activity may be thought by the producers and the public to be insignificant, but a huge number of small increments may add up to an unallowable hazard. For instance, an individual automobile or one small plant may pollute the atmosphere trivially, but we must think about the millions of cars and industrial exhaust chimneys. We should not assume companies will police themselves adequately, even though they may be expected to avoid undue dangers to their employees and customers and large lawsuit settlements. The community needs to be in the act. That, as we concluded in an earlier chapter, means government involvement.

Investigations, data gathering, assembly of the fundamentals of science and technology, and the study of statistics over long periods of time and large geographical areas require an organization of substantial size and scope, most often well beyond that of any single company. This presents us with a tough organizational problem of relationship between the private sector and government. We need regulation, we do not want to suffer from overregulation, and we do not want to curb useful innovation. So there is more to good organization than simply handing the job to government. But mainly we have done only that, assigned the government more and more regulatory duties as the SHEP problem has grown. The results so far have not been good.

Regulation of safety, health, and environmental protection has been costly, has delayed benefits, and too often has not given us the needed protection. The task has been enormous, so we should not have expected it to have been dealt with perfectly, nor can we expect perfection in the future. But things do not have to be as bad as they have been. We know better now, and we should regulate better. Sound innovation in the necessary regulatory process is possible and practical. If successful, then the gain for the nation will be great because while disbenefits are minimized, more advantageous technological advance will result. Without improvement, regulation will interfere with such advance and, worse, further expansion of technological activities too often will yield a net negative contribution. Few aspects of the management of our technological society are more important than recognizing the positive-versus-negative aspects of technological implementations, the permanent need of government involvement, and the handicap to progress which mishandling of the mission of regulation can create.

The Inherent Difficulty of Technological Regulation

Why is regulation of safety, health, and environmental protection so difficult? To begin with, defining accurately what hazards are tolerable is essentially impossible. The unwanted ills conceivably present are too numerous and not always quantifiable. Even if for every activity—use of fertilizers and pesticides, operation of autos and airplanes, running of nuclear reactors and coal mines—we could measure every possible menace, we still would not learn thereby what threshold level of impairment is acceptable. What we define as tolerable depends on how much we are willing to risk losing. How much lowering of our life expectancies or vigor or joy in natural surroundings are we willing to countenance? Since people differ in value judgments even when they agree on the facts, how can we specify the limit of harm? We cannot merely insist the disbenefit be negligible. If we did, how would we define negligible?

Just as identifying detriments is only a beginning to sound regulation, so is a listing of rewards. How much of a gain should we insist on before being willing to accept a given risk? Industry's dollar costs of meeting SHEP regulations are eventually paid by all of us, and most such costs can be estimated. But we cannot readily put an economic worth on improvement in health or prevention of accidents. No marketplace sets a price for one extra year of life or a month's supply of clean air. The sound decision making we ideally would like to achieve would include the step of listing the pluses and minuses for each alternative. But such lists will be perpetually incomplete, the items often unmeasured and unmeasurable. Nevertheless, we must make decisions, so we sit in the awkward position of pitting one unclear alternative against another. How are we to balance the gains against the risks with limited knowledge of each and no clear weighing scale?

The Value Judgment Factor

Value judgments have an overwhelming influence on all ideas about regulation for safety, health, and environmental protection. Our value judgments are inconsistent and emotional, characteristics inherent in the makeup of the human species. Consider that nuclear power expansion has essentially come to a halt in the United States because, despite the fact that in all the decades of its use no member of the public has been killed by a nuclear power accident, people worry greatly over the possibility of a super accident that might destroy 100,000 people at once. The famous Three Mile Island nuclear incident was disastrous economically, but the commission established by the president to investigate what happened concluded that no health impairment to the community occurred as a result of the accident, except possibly for the short-time psychological strain from increased fear that was felt by some living in the area. But 350,000 Americans die each year because of cigarette smoking. If they were to die all at once in one place, that would be viewed as more serious than when their deaths are spread throughout the year.

Similarly, only a small minority of American drivers fasten their seat belts, although there are 50,000 deaths and ten times that many injuries from the 30 million traffic accidents we experience annually in the United States. Again, a half million people, despite the readily available evidence that they are doing major damage to themselves, use heroin regularly. Many more millions use alcohol to excess and thereby cause their own premature deaths.

Tens of thousands of chemicals are being manufactured, and a thousand new ones are added every year. Billions of pounds of some are produced annually. Analyses have suggested that more of these substances might be hazardous than we have officially recognized. A hundred billion pounds of ethylene dichloride was created before the substance was found to be a strong carcinogen. Vinyl chloride was produced at a rate of 5 billion pounds a year before tests showed it to be cancer-causing. Recently the National Academy of Sciences stated that for pest control a billion pounds of toxic matter was being introduced yearly into the environment and that the government's knowledge of the potential harm was superficial. Over 2 billion pounds of hazardous waste, from explosives to cancer-causing matter, is moved about each year in 10,000 vehicles and stored or treated in 20,000 facilities, mainly without benefit of the federal policing called for in legislation already years old. The Environmental Protection Agency (EPA) frequently has missed deadlines for issuing regulations on landfills imposed by federal courts and has estimated that it will take several more years to issue most of the required permits. The EPA has completed examination of only a token number of the 50,000 chemicals required to be tested by the Toxic Substances Control Act of 1976.[1]

Air pollution is affecting crops, with potato yields down in Connecticut, spinach disappearing from vegetable farms near cities, and important grape

---

[1] The EPA, by 1980, had culled only thirty-eight chemicals as especially suspicious and deserving of testing, but the testing had not yet begun.

crops being abandoned in parts of California. Acid rain is impairing substantial areas of the continent. The regulatory agencies naturally interest themselves only in the specific hazards cited in the legislation that created the agencies. Thus, since the Clean Air Act contains no mention of acid rain, no agency is required to think of itself as having a specific responsibility to tackle that problem. Acid rain comes from power plant emissions that rise into the atmosphere and fall again to poison streams and lakes as far as hundreds of miles away. New York State believes it has hundreds of lifeless lakes and hundreds more in critical condition, and Canada is complaining about the acid rains stemming from industrial activities in the Ohio Valley. While there is no question that more research is needed on the causes and effects of acid rain, the National Academy of Sciences has concluded that the evidence tying acid rain to the effluents of power plants is convincing.

It has been estimated that 80 percent of all cancer is of environmental origin. If true, then in principle most such cancer should be preventable. One tough problem in labeling environmental carcinogens is that it may take twenty-five years before their influences are felt. This is true not only for many industrial chemicals but for the effects of low levels of radiation, radium, coal mine environments, asbestos, and numerous other factors. Instituting practical controls is often a problem, because of wide variations in personal values, even when serious hazards are positively identified. For example, many people continue to smoke (lung cancer), ingest dietary fat (colon and breast cancer), and enjoy charcoal-broiled steaks containing charred protein (a mutagen).

Shortcomings of Present Regulation

Clearly the task of identifying and measuring hazards is an enormous one. SHEP regulation is unsatisfactory partly because we have not faced up to the scientific and technological aspects of the task. Usually the technical experts and laboratory facilities required are more than government agencies are in any position to assemble. Inadequate budgets and powers of inspection often limit studies that could lead to good regulations and the policing needed to enforce them.

In the last twenty years numerous studies have indicated again and again that while it is relatively easy to obtain beginning scientific data to indicate that a risk may be present, it is usually extremely difficult to establish the level of the risk. Whether the subject of study has been the evil effects of carcinogens in water, cotton dust, saccharin, bacon, sulfur dioxide, or benzine, or the value of passive restraints for automobile occupants, the available data are almost always open to questions of interpretation. Makers of regulations face too broad a range of possibilities as they try to use available scientific evidence as a foundation for their regulations.

Over 300,000 government employees staff some fifty or more SHEP regulatory agencies. Examples are the Consumer Product Safety Commission (issues standards for more than 10,000 products); the Food and Drug Administration (inspects and tests drugs, cosmetics, and food products); the Environmental Protection Agency (sets standards and monitors discharges and emissions from polluters); the Occupational Safety and Health Administration (enforces rules for protecting employees in over 3 million workplaces); the Federal Aviation Administration (sets standards for airline maintenance, air traffic control, and pilot fitness); the National Highway Traffic Safety Administration (sets rules for fuel efficiency and safety standards for bumpers, seat belts, etc.); and the Nuclear Regulatory Commission (licenses nuclear power plants, sets standards for plant construction, supervises disposal of nuclear waste).

Critics of present technological regulation abound. Their complaints aver that regulation often does not provide the minimum protection needed; overregulation is frequent; Congress has passed bad regulatory legislation; the courts are called upon to do what they cannot and should not be asked to do; agencies sometimes have conflicts of interest; regulators often do inadequate investigating and stall to play it safe; value judgments are confused with economic or scientific factors; an unintegrated hodgepodge of disconnected decisions predominate; balanced decisions, with the risks and benefits of all real alternatives compared, are rarely made. Whether these criticisms are justified is itself a value judgment; my judgment is that all have considerable validity.

The EPA is way behind in the research and investigatory activities needed to provide a sound, scientific basis for its actions. It may well be that some present EPA standards are too severe, that there has been overregulation, and that this has been costly to the economy without necessarily providing additional health benefits in return for those excesses. At the same time, if EPA is to improve its regulations, reduce overregulation, speed permit granting, and accomplish minimum policing where the regulation is solid and necessary, it will need more research data.

The enforcement and monitoring functions of the EPA are as weak, compared with the need, as its research capability. The Toxic Substances Control Act and the Resources Conservation and Recovery Act were both passed in 1976 and relate to the identification, control, and disposal of toxic substances. It took four years for EPA to begin to issue regulations in response to these two acts.[2]

Meanwhile, most waste facilities are failing to comply with existing interim regulations, and because of EPA's inadequacy in staff, enforcement has had minor influence. It has been estimated that EPA would have to more than double its workload if by 1985 it is to be in a position to cope with the new toxic chemicals that are being poured into the environment every year.

When an agency overregulates, it is not necessarily the result of excessive zeal by the agency or its leadership's disregard of the potential negative impact of regulations on the economy. Overregulation may occur because the agency simply does not have enough data to issue superior standards, and its requirements may be severe because it feels an obligation to ensure safety at all costs. One way in which regulation can be made worse overall is to cut an agency's budget drastically, seeking thereby to make it impossible for that agency to overregulate. The EPA, understaffed already with regard to the task assigned to it, lacking both the research budget necessary to discover what makes good regulations and the investigations budget necessary to monitor the regulations when issued, is now faced with operating with a substantial reduction in funding.

When restricted in scope and speed of investigatory capability, while remaining anxious to protect against hazards, agencies sometimes hold back on approvals. Such delays restrict the potential harm in new developments, but they also may hold back the realization of valuable new benefits. Regulation as presently practiced frequently involves voluminous, costly documentation on minor issues and long negotiations. The bureaucracy required of industry grows to match the government's. Meanwhile, SHEP regulatory action sometimes appears to cause results exactly opposite to the intent. For instance, the current clean-air offset requirement mandates that "old pollution" has to be cut down before superior new plants can be built in the same area. Since it is not always practical to make an old plant pollute less, this rule means that up-to-date, efficient, low-pollution plants are discriminated against in an established region and will not be created there, while old plants that pollute heavily are allowed to remain. This discourages investment in the new technologies that underlie cleaner plants.

---

[2] The EPA has recently issued a 600-page book of standards for land disposal of hazardous chemicals. The Environmental Defense Fund has attacked the regulations as ignoring the significant toxic air pollution from these dumps.

It has become a common practice for regulatory agencies not only to insist on specified end results, but to mandate that the results be obtained in specific ways. Thus the government wants less sulfurous compounds to issue from the chimneys of plants burning coal. Instead of saying that the smokestack must be limited in its bad emissions to a certain level, the law requires that expensive stack gas scrubbers be installed even if, by choice of cleaner coal, the same emission standard could be met without adding the scrubber.

Some coal can be ground up and washed after it is mined, and if the washing is done properly, it will remove a quarter or more of the sulfur from the coal. This approach frequently would be cheaper and reduce pollution more than the use of scrubbers on the smoke generated by coal with higher sulfur. If the government agency would simply set down how much harmful material it would allow to exit the stack, the solution would be chosen by the private sector at the lowest investment and annual cost. It has been estimated that to comply fully with the scrubber rule would add some $5 billion to the industry's capital budgets for the decade of the 1980s. Naturally the effect is to bias the utilities against creating new plants, and the legislation actually encourages a continued flow of greater pollution from old plants.

Over a half million tons of lead is exhausted annually into the environment. Recent research strongly indicates that even minute quantities of lead can be very damaging to the intellectual capabilities and health of children, whose high metabolism makes them more susceptible to lead poisoning. Indeed, research indicates that an undernourished child absorbs even more lead. Specifically, some 5 percent of all preschool children and 25 percent of black children from poor families have been found to possess lead levels in their bloodstream above the present federal standard that seeks to end lead poisoning. All of us have in our bodies about 100 times more lead than did our ancestors. Yet in 1982 the government, seeking to avoid overregulation, seriously considered allowing oil refiners to use more lead in gasoline, just as federal funds for nutritional aid to poor school children were being reduced.

Several years ago the toxic chemical PBB was accidentally introduced into livestock feed in Michigan. It took almost a year for the error to surface as discoveries were made of the decreases in milk production, increases in sickness among milk cows, and a drop-off in egg production by diseased hens. During that year, the poison was spread far and wide through the distribution of farm products, even though tens of thousands of cattle and millions of chickens were destroyed. It has been estimated that some 97 percent of the residents of Michigan today retain in their bodies measurable residual levels of this chemical. It may take many years before it is clear whether these trace amounts will prove to be damaging, because carcinogenesis tends to lag for two or three decades. Before the accident it was known that PBB causes cancer in laboratory animals.

In most instances the most dangerous of wastes, including such toxic metals as arsenic and mercury, can be recycled; the remaining materials can be turned to safe ash or baked into insoluble lumps that cannot contaminate water supplies. If the government were to place heavy charges on each quantity of toxic waste dumped, the companies that produce them would find it economically sensible to pay for R&D and processing that would enable them to convert the waste into harmless material at least cost. This might be more beneficial to the national economy than having a regulatory agency create detailed requirements on how to dispose of waste.

Conflicts of Interest

Can a regulatory agency be an adequate investigator of negatives if it simultaneously has the responsibility to arrange a flow of positives? For instance, is the Nuclear Regulatory Commission in business to see that the nation develops and uses nuclear energy, or does it exist to protect us against the negatives of the nuclear approach? If both, does not the NRC have a built-in conflict? Moreover, is it reasonable to assume the NRC is in any position to judge what degree of hazards we should accept from nuclear reactors, unless it also is assigned the duty to determine how much energy the nation requires, is expert on alternatives like coal or solar energy, studies the politically acceptable level of dependence on foreign oil imports, and investigates the possibilities of a greater energy conservation effort? None of these other matters are mentioned in NRC's charter.

The NRC often has counseled the suppliers and utilities and until recently had never turned down a request for a license. When it also takes upon itself the roles of safety expert and public protector, then perplexity is natural. Recall the sudden frightening happenings at the Three Mile Island nuclear reactor installation, the disclosure of operating errors, the puzzlement as to exactly what had occurred and what risks were present, and the confused communications. When the crisis started, other utilities operating similar reactors elsewhere immediately considered halting operations as a precaution. The NRC was looked to for a decision on a nationwide nuclear reactor shutdown, a step that would have caused extreme inconvenience to industrial power users and homes in certain highly populated segments of the country already dependent on nuclear power. Here the mission of the NRC as a protector was understandably perceived by some to be in conflict with its role of ensuring uninterrupted electric power.

There is an opposite side to this conflict-of-interest coin. Those wishing to get on with technological activities are frequently frustrated by the negating activities and indecisiveness of government regulatory agencies and become loud detractors of them. Sometimes such critics of an agency are politically powerful, and the ultimate effect of their pressure is to cause the agency to depart from its mission of protection and seek a more even approach. In so doing, the agency automatically compromises its role as a guarantor of safety.

Consider, for example, the present political difficulties of the Occupational Safety and Health Administration. OSHA was not created either to promote industrial development or to slow it. The annual rate of Americans killed on the job has now reached almost 15,000, and over 2 million are disabled yearly by injuries. The cost to industry to increase safety in accordance with government rules has been estimated to be over $10 billion annually. However, the economic loss to the nation that results from the present rate of workplace accidents has been estimated to be even greater, more like $30 billion, suggesting that we have not necessarily reached the point of diminishing returns in efforts to seek economic gain by reducing the accident rate. Moreover, personal losses do not even figure into these economic ratios. We can put no economic value alongside the loss of life and the impact on the families of those who are killed or injured severely.

But industry has been disturbed for years over the costs of meeting OSHA's standards and the large staffs needed to interface with the agency. Some have suggested that the standards do not always enhance safety. Murray Weidenbaum, the first chairman of President Reagan's Council of Economic Advisors, has stated that according to the Labor Department's data the days lost per worker due to job-related illness or accidents is a flat horizontal line over the years since OSHA's standards have been enacted. It is, as Weidenbaum says, "as if OSHA didn't exist—no progress whatsoever despite all the rules—understand now: no improvement."

The criticism has become so great that OSHA is now on the defensive before Congress. It may have to create for itself a better reputation or face lower funding and restricted jurisdiction. It is worth noting that in 1981 the fines that it issued for serious violations dropped by 40 percent. At the same time, its actions will hardly be viewed by industry as contributing to economic expansion. Thus, all concerned—workers, industry, and government—can be expected to continue to be unhappy with OSHA.

Comparing Alternatives

Decision making on technological operations can hardly be sound unless it includes the steady examining of alternatives. There is no such thing as zero risk, so to seek it can only generate an expensive bureaucracy with no chance of succeeding. Comparing imperfect options, balancing their risks and gains, both in arriving at rules and policing their application, is key. If a regulation is overly severe, it is not necessarily an error on the safe side, because it could also have a negative impact on productivity and employment. It could hurt America's ability to compete in the world market. It could lower return on investment, raise prices, discourage new investment, and decrease average income. People who are made poorer because a weakened economy suspends their employment suffer from health problems just a surely as do normally healthy citizens whom we do not protect from health hazards. A university study of thousands of cases showed that middle-class people, after a single job termination or personal financial setback, suffer twice the rate of illness and accidental injury.

Starting with the Pure Food and Drug Act of 1906, we have added laws governing therapeutic drugs, cosmetics, medical devices, occupational environments, pesticides, children's sleepwear, automotive safety, nuclear emissions, pollutants in water and the atmosphere, and more. Each piece of regulatory legislation has been narrowly focused, and the various legislative steps have all been disconnected. Effects of regulations in one area on other government programs, on the national economy, and on public health would enter the deliberations if a responsibility to compare alternatives always accompanied government regulation. The closest approach to this desired balancing is that in some instances agency charters call for finding evidence of positive benefits before allowing a new product on the market. However, the legislation setting up SHEP regulating agencies is usually silent on defining trade-off duties. In fact, Congress sometimes has actually forbidden balanced decisions by the agencies and required unbalanced ones. The Clean Air Act specifically precludes the deliberate weighing of benefits versus harms. The Delaney Amendment to the Food and Drug Act tells the FDA that it must not consider the cost impact when constructing regulations.

The Clean Air Act is dominated by the absolutist nature of the mandated ambient air quality standards which are required by the law to be set at a level that protects the health of the most sensitive groups of the population, apparently without regard to costs. Since the present status of science does not allow us to relate the effect to the cause at low levels of pollutants, the only way the standards of the Clean Air Act can be met is by arranging for zero pollution, a decidedly theoretical requirement since it cannot be implemented.

Even where the Clean Air Act does not have to be interpreted as asking for perfection, attempts at strict adherence to it by the EPA, it has been estimated, would require some $300 billion of expenditures during the 1980s. Of course, this would be borne by the public. It would raise prices for U.S. industrial output (compared with foreign-based output) and would affect our economy adversely through increased unemployment. During this same decade some of the most populous areas of the country would have to halt all further construction and growth because they would not be able to meet the Clean Air Act's requirements unless they were to shut down old operations whenever they add a new one. But then the subsequent mandatory write-offs of previous investments would add up to more than the industries' capital structure would allow.

Two decades ago it took five years and $1 million to work an average new pharmaceutical drug through the regulatory mill. Today the typical cost is nearer $20 million and the time exceeds ten years. In this period the rate of new drug introductions by U.S. firms has fallen by 50 percent. The costs of R&D in drugs have become so high that smaller U.S. companies have essentially left the field of discovering and developing new drugs. Perhaps we are being wise in paying out time and money to prevent hasty introductions of harmful drugs. Or are we allowing needless suffering and deaths that new good drugs might prevent? We don't know, because no group has the assigned function of answering such questions of comparative trade-offs.

The increased severity of drug regulation in the United States has greatly decreased the chances of dangerous drugs appearing on the market. However, the higher investment requirements, the longer time period before a return can be expected, and the shorter beneficial patent protection period (because so many of the seventeen years granted in a patent are used up in the approval process) have led to a marked lessening of innovative drug development in the United States. It has been estimated that of 10,000 new drugs that are begun in R&D, only 1,000 survive through animal laboratory testing, ten are tested in humans, and one survives to be marketed. New surgical and diagnostic techniques are important in medical advances, but it is generally accepted that the greatest contributor to improvement in the treatment of disease is new drugs. Thus, when the environment for innovation becomes highly unfavorable, it is not only that the availability of valuable drugs is delayed; other beneficial drugs simply are not even started in development. It is to be noted, of course, that it is not the FDA's job to be concerned with creating motivation for innovation.

During the 1970s, U.S. drug companies increased their annual R&D budgets in foreign countries from under $50 million to over $250 million. Trials with volunteers are permitted by other countries, who view differently the balance between the dangers of new drugs and the benefits they might provide. If pharmaceutical R&D moves abroad, then foreign countries, not the United States, will be penalized by the hazards, but they will be earlier beneficiaries of the health benefits and will reap the major financial returns. Perhaps it works out that this pattern affords us high protection and costs us little in missed gains. If so, that would be largely accidental, because no agency is clearly charged with comparing these broad alternatives.

The Department of Agriculture estimates that if all pesticides were banned, crops would decline 30 percent and food prices would rise 75 percent. Millions of people around the world would go hungry because U.S. food would no longer be available to them. Unregulated use of pesticides is unthinkable, but the standards obviously should not be based alone on the dangers of their use but also on the disbenefit of their nonuse, one alternative pitted against the other. A recently introduced herbicide, said to be much superior environmentally to existing ones, was the result of twenty years spent in research and in the approval process. Should the regulatory parade be accelerated to realize earlier gains in the expectation that they will exceed the harms of premature approvals? To weigh probable benefits against risks is not now a required standard procedure.

The trade-off between improving the environment and increasing the energy supply is typical. If coal use is expanded, then energy supply will be enhanced but SHEP hazards will increase. Letting the economy slow down because of a dwindling energy supply is bad. Allowing more pollution and accidents is also bad. Balancing the positives and negatives is mandatory. However, in unrelated acts, the government first imposed drastic controls on coal use. To cut air pollution, it mandated that utilities using coal change over to oil and gas. A little later, reacting to OPEC actions, it decreed greater use of coal. Meanwhile, with no one in charge of comparing alternatives and balancing the positives and negatives, the government set a low ceiling price on natural gas. This simultaneously increased demand and discouraged further exploration. The ceiling price was kept on even though double-digit inflation arrived and greatly magnified the mismatch. The government energy policy preached conservation but encouraged dissipation (by keeping conventional fuel prices low). Then, having made development of new domestic energy sources through private investment less attractive, it started government-funded programs to pursue new energy alternatives.

SHEP regulation is often self-contradictory and violates common sense when it fails to consider the inevitable impact of a ruling on the rest of the economy. The automobile pollution problem is a strong example of this because the automotive industry employs more people, constitutes a higher fraction of the nation's GNP, uses up more materials, consumes more energy, and is more influential on our way of life than any other industry. The hoard of automobiles in the United States has been called our second population, the inventory of vehicles being roughly equal to the human population. Government regulations affecting the design and price of a car have an enormous effect on unemployment, the overall national economy, our international competitiveness and, because of air pollution and accidents, the health and safety of Americans. The price of a car influences the number of cars produced and sold and the rate at which the public shifts from older cars (representing lower MPG, more pollution, and less safety) to more desirable ones. Government actions dominate manufacturers' decisions as to where to put available funds and whether to meet regulations, enhance productivity, or improve the product.

How has the government been handling regulation relating to matters automotive? First, it introduced strong air pollution restrictions on automobiles without considering the impending oil shortage. The EPA's isolated auto emission rules mandated unleaded gasoline and lowered MPG performance, thus increasing the demand for petroleum. Less gasoline is produced from a barrel of crude in making unleaded fuel, so more refinery input was needed, creating additional pollution.

Seat belts, safety glass, collision-proof door latches, and the energy-absorbent steering column constituted the first mandatory safety requirements. The regulatory bureaucracy then invented 5-mph bumpers, the airbag, and the interlocks of seat belts with the ignition. The public vetoed the last two. The new bumpers have perhaps reduced repair bills after some accidents, but have cost consumers a billion dollars for the adornment and have required hundreds of millions of gallons of extra gasoline annually to handle the added weight. It is not evident that any safety benefit has been attained.

Serious consideration is now being given to easing the current emission standards for automobiles, even though these standards are being met by cars presently produced. Doubling the permissible levels of carbon monoxide and nitrogen emissions is proposed. This would reduce the price of a car by around $50. The penalty that might be paid, while saving this $50, is not clear because the research results on the relationship between health in the long term and the extent of air pollution are as yet inadequate.

Diesel passenger cars are being encouraged because they use less fuel and because the fuel can be produced with less crude oil. However, diesel-powered vehicles produce some 50 percent more soot or particulates than gasoline-powered cars, and there is confusion about what emission standards to set for diesels. So far the scientific data are highly incomplete, since tests on animals to date have not pinpointed a serious problem traceable to diesel exhaust, even though that exhaust darkens the air. The technology does not yet exist to meet standards more severe than existing ones. The best methods currently known to minimize soot unfortunately increase an engine's output of nitrogen oxide, which combines with the inherent products of diesel combustion to produce compounds definitely known to be more hazardous than soot.

Thus it is too early to evaluate the health hazards of a very large increase in the number of diesel passenger cars driven in America. However, if things go as they have in the past, the EPA might increase the severity of standards before the scientific data are available to justify it, virtually eliminating diesel passenger cars from the American market. This would increase the amount of gasoline burned, with its train of negative consequences, from pollution from the refining and burning of gasoline to the social and economic disbenefits of further dependence on imported oil. Such a pessimistic possibility is not to be ruled out unless we find a way to handle broader trade-off considerations, rather than isolated and narrow ones, when automobile regulations are at issue.

To this day, we do not know whether government auto standards as to safety, fuel economy, and emissions are in the right range, everything considered—unemployment, trade imbalance, oil supplies, and inflation, as well as the disbenefits of pollution and casualties the regulations were intended to reduce. We do know for sure that the rules, in origin and execution, were isolated, not integrated.

Our clumsiness and inadequacies in SHEP regulatory decision making, when the key issue is comparing alternatives sensibly, are illustrated well by a great controversy over Clean Air Act amendments now being proposed. Billions of dollars are involved in estimates made by the two sides, one that wishes to weaken the standards for clean air called for in the act, and the other that wishes to retain them or make them even more severe. Thus some estimates put the costs to date of complying with the Clean Air Act at around $15 billion. On the other hand, other estimates rate the benefits realized from reduction in air pollution since the act at over $20 billion. The costs are to consumers in higher prices or higher taxes, and the benefits are in greater economic value of crops and decreased damage to property. The economic estimates, of course, do not put any value on lives saved. There have been estimates (not universally accepted as accurate) of the number of deaths associated with air pollution each year at between 100,000 and 200,000. Even more important, in trying to make sense out of these numbers, we do not know how either the number of deaths or the amount of money spent or saved would be influenced by raising or lowering the standards when all of the interrelated impacts are introduced into the equation.

For example, making the standards more severe might not actually lower the death rate, either because the incremental improvement, starting from the situation we have today, would be too small or because increased unemployment and lower general health might create more illness as car sales fell off. What we have done to clean up the air in America has given us human benefits for sure. This does not, however, tell us where we are in the trade-off between potential further benefits and the disbenefits that would come from making the standards more severe.

Let us suppose that we spend an additional billion dollars to reduce auto emissions and in so doing we save a thousand lives. These would be people whose sensitivity to air pollution (compounded by their existing diseases, such as emphysema, asthma, or cancer) would cause them to be pushed over the line unless the air they breathe is improved. But if we are to assign $1 billion to decreasing unnecessary deaths, there are many other quite different programs to consider as alternatives. Perhaps we could save more than 1,000 lives in the next decade by using that same billion dollars in a campaign to lower the popularly accepted practice of driving under the influence of alcohol.

With Congress decreeing that agencies set rules to eliminate hazards to the maximum extent, we certainly should not be surprised if some expenditures save few lives compared with what could be accomplished if the same funds were invested in other areas where hazards are more deserving of the attention. When fragmented, vague, and perfectionist guidance is given to the separate agencies, little comparative data are produced to settle such questions.

Improving the availability and the speed of paramedic service to decrease heart-attack fatalities should be compared with improving the lighting on highways to cut down on automobile fatalities. Or, as another example, what would the results show if a study were made of longevity and medical costs per year of adult life, each related to the quality of nutrition the individual had received during childhood? It would not be amazing if the cost of providing good nutrition to each child (who does not now receive it) was found to be substantially less than the cost of later medical care traceable to poor nutrition. We have no group in charge of making these comparisons and influencing the budgeting choices of Congress so as to put available funds in those areas that appear most deserving, those that yield the greatest saving of lives for the money spent.[3]

At a time when the nation is severely pressed economically to provide for all that the citizenry expects from the government—including a strong national defense program, tax reduction incentives for a growing economy, and a balanced budget—almost $15 billion a year is spent on smoking-related health problems of cancer, heart disease, and emphysema, and the associated lost production and wages involve another $25 billion. Yet Congress and the executive branch, each working feverishly to cut a billion here and a billion there in order to attack a $200 billion deficit, seem relatively uninterested in mounting an effective government program to bring down the shameful number of smokers (one-third of all Americans). It is not an amusing fact that the U.S. government spends well over a billion dollars on cancer research while the industry that produces the product government research labels as the number-one cause of cancer spends even more to persuade the public to smoke. (The government budgets a few million dollars a year to advertise the hazards of smoking.) One can be forgiven for wondering what a billion-dollar program of hard communication on the dangers of smoking, designed for young people, would accomplish to cut cancer deaths of Americans. If this expenditure is out of order for the government, then why should it spend money to investigate cancer caused by smoking in the first place?

Few records of government regulatory activities can match—for confusion, illogic, inadequacy, and sheer randomness of decision making—that of the U.S. government in handling the licensing and control of nuclear reactors to generate electric power. Let us take a very recent example first. The present administration, in its rhetoric both during the campaign and afterwards, attributed the nation's nuclear power problems to pressure from antinuclear activists. It announced expedited licensing as a policy to get the nuclear reactor program rolling again. By mid-1981 the administration had appointed two new members to the Nuclear Regulatory Commission, including a new chairman. But within a short time the commission revoked the start-up license for California's Diablo Canyon nuclear reactor, and the new chairman was very critical of the nuclear industry's quality assurance programs. At about the same time, a dozen partially completed reactors were canceled on economic grounds. No new plants have been ordered. The

---

[3] Increasing lead in gasoline to save energy costs will undoubtedly create more brain damage and retardation in poor children, putting an excessive additional burden on society to care for them later.

government's highly pro-nuclear energy policies seemed to scare Wall Street as much as the anti-nuclear propagandists.

A sensible nuclear regulatory policy for the nation would encompass adequate reactor safety research, keeping nuclear reactors for peaceful electric power generating purposes well separated from the issue of bomb proliferation, getting on with long-term waste disposal implementations, and arranging appropriate integration of nuclear reactor policy with overall national energy policy, itself a component of overall national economic and defense policy. In all these matters the government's performance ranges from weak to neglectful. Instead of dedicated action we have nuclear policies that merely sound like political campaign propaganda, pro-nuclear power in an isolated, absolute sense.

Shifting our look at nuclear policies back to the past, we note that the government built its regulatory system while forecasting that there would be 1,000 nuclear reactors by the year 2000. This would have required construction permits and, later, operating licenses issued at the rate of two or three a week. The regulatory apparatus that was set up could not possibly work out the criteria and make the detailed analyses of proposed projects required to keep up with a schedule this ambitious. The whole nuclear reactor business has come to a standstill because even the modest rate of growth of nuclear reactor activities during the 1960s and 1970s turned out to be far greater than either industry or government could manage and regulate. All the while, the government has not done what would appear to be the most basic task of all. That is to evaluate how much of a contribution we need from the nuclear dimension in view of the demand for and the supply of oil, coal, gas, and shale, the impact of conservation measures, lower-MPG automobiles, and the rest of the stream of considerations and alternatives that make up a national energy policy.

A prime example of regulatory malfeasance by the government is its failure to handle the problem of nuclear waste disposal. Without a single additional nuclear power plant in the United States, 100,000 tons of radioactive waste will be sitting in ponds outside the existing plants before this century is over if nothing is done.[4] Some existing reactor installations will run out of room for storing their used fuel rods and will have to stop producing power well before the end of this decade.

---

[4] Near Buffalo, 560,000 gallons of poisonous waste is in a carbon steel tank in a concrete vault. The radioactivity will be dangerous for thousands of years, and the tank's life is estimated at forty years. The problem has been the subject of negotiation for years without agreement on a plan for more permanent disposal.

We have had the technological capability for years to dispose safely of the dangerous waste produced by nuclear reactors that generate electric power. One of the reasons the federal government has delayed the initiation of a permanent waste storage program is that most states found to possess the stable underground structures needed to constitute suitable sites raise objections to storage in their areas. But here is an example where innovation is needed, and not along technological lines. If the federal government held an auction and indicated it would locate the storage facilities in those states that offer to take on the assignment on receipt of federal money, then bids would undoubtedly come in. For the right price, almost any state would accept the facility. Storing the waste is a component of the economics of utilities electing to take the nuclear route, so it is reasonable to expect that each utility producing waste should be required to pay for storage in proportion to the amount of waste produced.

For decades a fundamental part of the U.S. policy of nonproliferation of nuclear weapons has included a careful separation of military and civilian nuclear programs. Now, because the Defense Department is planning new nuclear weapons for the decade ahead, a large increase is required in the rate of production of plutonium enriched to bomb grade. Most existing nuclear bombs use uranium 235, but the new weapons planned for the cruise missiles, neutron weapons, Trident and MX missiles, and others will be designed around plutonium because this leads to reductions in overall size. Thus the retiring older bombs will not provide the required amount of plutonium, and the nation needs to step up plutonium production. A large quantity of plutonium is contained in the spent fuel rods of nuclear power reactors. Extracting this plutonium may solve two requirements at once, obtaining bomb-type plutonium most economically and reducing the severity of the nuclear waste problem.

However, the fundamental idea that nuclear materials and facilities developed for civilian purposes should never be used for military purposes is at the heart of the nuclear nonproliferation treaty. Abandonment of this principle by the United States might cause the entire nonproliferation campaign of the last few decades to fall apart, because every nation in the world would then feel free to engage in converting materials from its civilian reactors to bomb-grade levels. Making the decision on which way to go in filling the plutonium requirement is moving as slowly as the handling of the nuclear fuel disposal problem.

As a final example of the obviously needed, but rarely provided, comparative analyses and judgments in regulatory decisions, consider the problem of formaldehyde, a component of many building materials, including plywood and foam insulation. Because tests have shown that in very high concentrations formaldehyde causes cancer in rats, the Toxic Substances Control Act and the charters of the Consumer Product Safety Commission, the Occupational Safety and Health Administration, and the Environmental Protection Administration have given each of these agencies ample reason to consider banning formaldehyde products in buildings. However, scientists are not at all convinced that formaldehyde really poses a cancer risk at the low levels of human exposure in typical buildings.

While none of the agencies have been given the mission of considering the economic impact of a ban, it should be noted that formaldehyde is an ingredient of importance in around 10 percent of the nation's GNP. This means that there would be severe economic repercussions in a total ban; impact on the health of the citizenry through heightened unemployment might exceed that of allowing continued use of formaldehyde as a building material. Only with extensive additional testing can the risk be made known. But even should the hazard be evaluated, the government's administrative apparatus will not lead us automatically to the best trade-offs or to regulations on the use of formaldehyde in which all impacts have been properly considered relative to one another.

Who's in Charge?

Aside from failing to compare alternatives, the nation's regulatory organizations leave great gaps in decision making. For instance, the eastern United States has a refinery capacity for less than a quarter of the oil consumed there. No new refinery has been built on the east coast for over twenty years, and petroleum products must be shipped from a distance, using energy and adding pollution from its dissipation. Those seeking to locate a new refinery on the east coast have contested for many years with those striving to prevent it. Involved, in addition to those who would operate the facility, are numerous citizen groups, the Department of Energy, the Department of Interior, the National Oceanic and Atmospheric Administration, the Commerce Department, local government groups, the Army, the General Accounting Office of Congress, the Environmental Protection Agency, the Coast Guard, and others. None has decision power.

To secure approval for a California-to-Texas pipeline, the Sohio Company spent five years obtaining 700 separate permits from numerous regulatory authorities. Then, seeing no end of legal challenges, the company gave up. But perhaps we badly need the pipeline. Who knows, and who is to say?

A different kind of example is the Georges Bank project off the coast of Massachusetts where the Labrador Current and the Gulf Stream converge and stir up nutrients. The fish catch there during the remainder of the century could be worth $3 billion to $4 billion. Geologists estimate that during the same period $10 billion of oil and gas could be obtained from the area. The government is selling petroleum leases amidst controversy over potential harm to the fishing. Granted that dollar values are only one factor, which product supply is more important to the nation, oil or fish? Maybe, as claimed by some, the oil searches and removal won't hurt the fishing. No one is charged with answering this question. Many agencies are involved, but no one of them has the responsibility to make the decision.

The Role of the Judiciary

In view of the value-judgment problem, the isolated actions of the many agencies whose rulings overlap, the lack of comparative analyses, and the indefiniteness of who is in charge, interested parties now commonly seek relief from regulatory edicts through the courts. Litigation has become so frequent that regulations are often rendered academic, since their enforcement requires a judgment in court. Consider an example.

The U.S. Court of Appeals, later upheld by the U.S. Supreme Court, recently struck down a regulation by OSHA on the handling of benzene. Benzene, like gasoline, is a very useful but generally dangerous and potentially explosive chemical. Breathing it in substantial concentrations makes one ill, and if inhaled over a prolonged period, it may cause cancer. Regulations originating several decades ago limited the allowable molecular concentration in industrial establishments to 100 parts per million. This was later lowered to 10 parts per million. More recently OSHA decreed that the molecular concentration should be decreased by a factor of ten, to one part per million.

Would adhering to these more severe standards save 100 lives annually? Even one life? OSHA had not performed tests to answer such questions, but it acted with conviction based on experience with other carcinogens. On the other hand, it was quickly ascertained that OSHA's new requirements would lead to industry expenditures of over $500 million. Immediately, then, a value issue arose: certain and large economic penalties versus some uncertain, perhaps negligible, health benefits. Who should decide this question?

Benzene is always present, if we consider occasional molecules as a presence, because it is put into the air by trees as well as by petroleum and coal. OSHA simply was going on the assumption that if holding benzene in the work environment to a low value is a good idea, then reducing it to a lower value must be a better idea. It assumed that the level of industry spending to meet regulations is not to be considered when the agency seeks to protect human lives. But the court ruled that at a certain level price becomes prohibitive, that some of the expected benefits must be measurable, and that OSHA could not apply the more severe standards.

A decision concerning a technological hazard is not merely a matter of performing a scientifically based risk assessment. If it were, we could turn the decision over to scientists. Broad problems arising out of technological advance are never matters of science alone. Instead, as with all other important issues, they are dominated by their social, ethical, and political dimensions. Thus some regulatory decisions are bound to end up in the courts.

The inadequacies of the regulatory process, while making the role of the courts more important, have also caused their function to be less distinct. Industry often complains that the courts unduly delay and interfere with industrial development. Accusations are common that judges, without adequate knowledge of the highly technical matters involved, misuse their injunctive power and are available at the beck and call of environmentalists. Labor and environmentalists, on the other hand, argue that the courts defer valid policing by regulatory agencies and are too subservient to industry. Thus it was an industry-supported public-interest law firm which led the charge into court to invalidate California's statutory regulation of nuclear power. When antinuclear demonstrators threatened to occupy the site of a nuclear facility which they claimed posed a clear and present danger to their community, industry did not hesitate to ask the courts for injunctions giving property rights a higher legal position over the right to interfere with possibly dangerous operations.

Congress's Defaults

While fundamental questions exist as to the appropriate role of the judiciary in technological regulation in our democracy, weaknesses in the legislation produced by Congress are in any case a key factor in causing actions to come before the courts. It was once the plan that Congress would define SHEP regulatory policy. Then, to carry out its policy, Congress would set up a regulatory agency to write detailed rules and enforce compliance. For years, if anything has become clear, it is not the policies set by the Congress. It is rather that this division of function—between setting policy and executing it—works poorly. Distinguishing in practice between policy and execution has not been as easy as once thought. Congress inadvertently gave regulators unclear but enormous power with insufficient policy guidance.

Congress sets up a new agency almost every time a new harm surfaces, and the empowering legislation often seems to direct the agency to do something bordering on the impossible, such as essentially eliminating the identified risk. The laws governing the agencies do not tell them whether to tolerate an insignificant risk if the cost of lowering it is huge or if banning the activity that gives rise to the risk may deny us a significant benefit. Most often when Congress has created a regulatory agency, that action has been inadequate because the agency is put together in an attempt to satisfy those constituencies that want it and to minimize antagonizing the constituencies that oppose it.

By creating many narrow agencies and ignoring the impact of each separate agency's regulations on the rest of the nation's affairs, Congress has not met its constitutional role as overall policy maker. Instead it has created the foundation for the resulting piecemeal and inconsistent regulatory actions. If Congress did its job properly and created adequate organizations for handling technological regulation, then the role of the courts would become less cloudy and more productive. Some have suggested that the cure is for Congress to engage in vetoing specific regulations that it does not like. But we know that this would simply put on Congress the burden of trying to satisfy its various constituents as they come in to complain about a regulation. Anyway, Congress has too little time to engage in this kind of detailed effort. Something different needs to be done to improve regulation.

All important decisions affecting American life are in the end political, because for stable acceptance and progress they must reflect the value judgments of the voters. Congress obviously deals constantly with sorting and prioritizing the value judgments of the public, and Congress is the natural body to figure out and respond to the will of the majority of the citizens. Congress, not the courts, should furnish the arena for the complex interplay among competing political forces. Yet this has not been reflected in the way the agencies have been organized. In actuality, when the agencies have been set up, no mechanism has been included to tap value judgments and use them to guide in decision making. Moreover, Congress has not helped with its pork barrel programs in the environmental field. For example, it has often insisted on sewage treatment grants to local municipalities to pay for facilities whose costs far exceed their benefits.

Most regulatory agencies hardly even possess an effective legal foundation for their control power. The agencies' heads do not answer to the voters in any meaningful way, and each new empire has been constructed to act largely independently of elected officials. Congress has committees to oversee the work of regulatory agencies, but these committees seem not to spot overregulation or underregulation readily, or the lack of motivation or responsibility of an agency to compare alternatives (if that is what the Congress intended the agency to do) or the inability of an agency to obtain essential facts basic to sound regulation because of inadequate staffing and budgets. Congress's performance as overseer is eager and timely only when it is looking for a culprit or a scapegoat after a conspicuous regulatory boner surfaces publicly. Members of Congress running for reelection can make little use of a record that shows they applied themselves earnestly and wisely to overseeing a regulatory agency.[5]

The amount spent per year by industry to meet government regulations is not much below industry's yearly capital investment or the annual federal tax revenues from business. Thus Congress could be excused for spending almost as much time studying costs of regulatory operations, to make sure they are sensible, as it does pondering taxes.

Hampering Technological Innovation

Overly negative regulation or poorly managed regulation of technological activities by the government is a severe handicap to beneficial technological advance. This stimulates some influential groups, aware of this handicap, to criticize regulatory practices. Also, some excessive criticism of government regulation is bound to come from entities in private industry merely because they would like to be free of the burden of meeting government rules. They can easily justify their views by citing the regularly available individual cases of ceaseless delay and apparently bad rules and the incompetence of the

---

[5] In a 1974 speech, former FDA administrator A. Schmidt, stated, "In all of FDA's history, I am unable to find a single instance where a Congressional committee investigated the *failure* of FDA to approve a new drug. But the times when hearings have been held to criticize our approval of new drugs have been so frequent that we aren't able to count them. . . . Whenever a controversy over a new drug is resolved by its approval, the Agency and the individuals involved likely will be investigated. Whenever such a drug is disapproved, no inquiry will be made. The Congressional pressure for our negative action on new drug applications is, therefore, intense."

bureaucracy that makes the regulations and polices them. However, no matter how we might restructure government regulation of technological operations, we shall still have both under- and overregulation in important areas and ample evidence of waste and ineptness—bureaucratic, technical, and other—because the task is so difficult. If the most outstanding brains in the country were readily available to the agencies, they still would not score well on tasks involving so complex a mix of science, technology, economics, politics, and personal value judgments.

Unfortunately the regulatory process ties up technological resources. If these resources could be assigned instead to the increased development of new advantageous products or the improvement of productivity or the conservation of energy, the nation's standard of living would rise. Then the national government would find it easier to provide for both a sound and growing economy and for national security. The costs to both the government and the private sector of all aspects of the regulation of technological activities are difficult to assess with precision because there are gray areas of definition, but such costs have been estimated at over $100 billion annually. This covers the investment cost of modifying facilities to cut SHEP effects, the research and development expenses to understand those effects, the added costs of redesigning products and the means of manufacturing them, and the heightened charges to consumers of articles whose prices have been increased to cover the costs of meeting government regulations.

This $100 billion is a large number, larger than the total of all annual expenditures for every kind of research and development by both the federal government and private industry. Still, we would accept this enormous expenditure if it achieved its real purpose. Unfortunately, a significant portion of what is being done to meet regulations today is not really worthwhile, is misdirected, or while intending to eliminate disbenefits, fails to do so. Costly effort is spent on meeting misconceived standards. Recovering some of this funding and applying it to beneficial efforts instead could make an important difference in the nation's economic progress.

Today, whenever a new technological operation is being considered or a new application of recent scientific discoveries is pondered as a source of new products, the possibility of incurring risks from unseen negatives puts a damper on the investment process. The negatives of a new activity cannot all be assessed ahead of time. Some disbenefits will only be apparent after the product is in use for years. It is even more difficult to guess what the government's regulatory apparatus will decree about the hazard. Even if the product is worthy of being deemed safe, because the few risks have been taken care of relatively easily through careful design of the product, the entrepreneur-investor cannot assume that the product will be perceived as safe by either the public or the government. Delays and expensive design changes, merited or not, may become mandatory by law, and this will add greatly to the investment and to the time lag from the inception of technological effort to the hoped-for returns on the investment.

Anticipating overreaction to and overregulation of potential hazards makes some participants in new technology nervous and cautious. Like inflation and high interest rates, the archenemies of technological innovation discussed earlier, overregulation acts to influence management toward conservative incremental steps and away from bold and speculative projects. Since most other countries view the regulation problem quite differently and appear to be willing to accept greater SHEP risks, our policies handicap the United States in international competition.

Innovating to Improve
SHEP Regulation

In view of the basic nature of technological regulation, let us consider a novel regulatory organizational scheme that is possibly superior to present approaches in timeliness of action, effective reaping of benefits from technological advance while protecting against hazards, and minimizing court actions on items best handled through administration and legislation.

Improvement starts by decisively separating the two basic duties of the government in regulating technology: (1) to investigate and provide knowledge concerning negatives; (2) to balance the good against the bad of various alternatives and make regulatory rules.

We start with a competent organization to uncover, research, study, assess, and provide recommendations regarding all hazards to safety, health, and the environment. To remove present conflicts of interest and confusion as to mission, we would relieve this investigatory unit of all responsibility (or even the slightest appearance of it) to consider positives as well as negatives and attempt balanced decisions. Decision making and regulating would not be its business. When it comes to earth, air, and water pollution, nuclear radiation, health hazards from toxic chemicals, occupational safety and health, purity in food and drugs, and all the rest, this group would be equipped with the required experts, tools, facilities, and budgets to enable it to track down detriments in existing or proposed activities with a required reasonable depth and thoroughness.

The operations of the Federal Bureau of Investigation constitute a useful analogy. The FBI is an investigatory agency. It investigates crime and finds criminals. It does not try or sentence lawbreakers. It does not issue edicts on what is to be considered a crime. It does not ponder whether capital punishment is proper or whether jails exist to punish or to rehabilitate. When it finds culprits, it turns them over (along with the evidence it has found in its investigations) to another part of the government.

If we accept the value of a separate, unambiguous mission to investigate disbenefits, it seems eminently sensible to bring all such activities together in one agency. Every potential harm to humans and the environment requires for discovery and evaluation an array of measuring equipment, laboratory facilities, and field offices and expertise in chemistry, physics, biology, engineering, toxicology, statistics, and other disciplines. Efficiency, synergism, and advantageous flexibility of organization would result if the experts and tools were all in one strong unit. Need would no longer exist for Congress continually to discover a new danger and launch still another new agency to investigate it. All investigation of existing or potential dangers from technological activities, known or unknown, would be housed in this single SHEP investigatory agency.

Let us suppose that the new, broadened investigatory unit has been created and does its job properly. We now also need a SHEP regulations board. Its task will be to make decisions in the form of rules and regulations, to issue licenses or permits as required by Congress or deemed necessary by the board, and to ensure compliance with its standards by fines or removal of licenses. The board, unlike the investigatory agency, would have the role of comparing alternatives, balancing the good against the bad, and the duty to connect each case that it considers to other national interests. It would have the unquestioned responsibility for banning or approving the challenged technological operation.

The legislation setting up the board would provide an approximate definition of the board's jurisdiction and would name those activities (such as nuclear reactor installations) specifically requiring approval and licensing by the board. More generally, Congress would expect the board to take on the job of identifying hazards that require federal controls and then go about issuing suitable regulations. Sometimes the board might not act quickly enough to satisfy Congress, and legislation might prompt board action. At other times, the board might be seen by Congress as overdoing regulation, and Congress would then step in to mandate that the board be satisfied with lower standards or easier-to-meet criteria for a license.

Estimating a risk would be a task suitable for the investigatory agency, but deciding what should be an acceptable level of risk is a sociological, political, or value question, and it would be a matter for the regulations board. The board would want to avail itself of the best estimates of risks and gains, but would then go on from there to consider numerous other issues and judgments that eventually would lead to its decisions. As an example, scientific studies have shown that synthetic fluorocarbon gases used in spray cans cut down the protective ozone layer in the stratosphere and that this reduction might result in increased skin cancer. On the other hand, it is known that a substantial number of people choose to develop a suntan and perhaps receive more harmful radiation in the course of a year from their sunbathing habit than from the depleted ozone. Perhaps, then, it is sufficient that the public be made award of the risks. At any rate, how to handle this issue would be for the board to decide.

The board would rely heavily on the investigatory agency to call its attention to potential hazards deserving board action. The board would expect to receive recommendations for actions from the investigatory agency. However, the board would not be dependent on the investigatory agency alone for its facts and advice. It would be expected to seek out expertise from the nation at large. The president, Congress, and the investigatory agency would each be empowered to call upon the board and deposit SHEP issues with it. Also, other private or public groups or individuals might request the board to consider certain problems.

The board would consist of several presidential appointees. Reporting to the board would be a substantial career staff organized into specialized groups—nuclear energy, drugs, air pollution, etc. The board would also have a general systems staff that would concern itself with the interactions of existing and proposed regulations. This systems group would study alternatives and make trade-off analyses among them. Much of the detail of such analyses might be provided by outside groups hired specifically to make individual studies. The board's responsibilities would include transcending the separate, narrow regulations typical of today's regulatory agencies and considering the overall impact on the nation of the various regulatory standards it issues. Thus rules affecting automobiles would be examined with the goal of understanding how standards on such separate issues as MPG, air pollution, and safety all interact, and how all these influence the health of the automobile industry, unemployment and family hardship, and the overall economy of the nation.

It is reasonable to ask whether the present system—many separate organizations specializing in various hazards, such as drugs, nuclear reactors, etc.—is not more efficient, precisely because of that specialization, than a system in which all the decision making is centered in one unit. The answer is, bluntly, that we can easily kid ourselves about the potential efficiency of today's approach. The single board proposed here would have its separate divisions, each busily engaged in the details of its own specialized area, even as occurs in today's independent agencies. The divisions could do their narrow work with the same opportunity for efficiency through specialization as is true of the many independent units in the present system. But today's units are not integrated. They receive no coordinated supervision that recognizes that their separate decisions inevitably interact to affect the overall national interest. In our new proposal, we have recognized—through the single decision board to which all the specialized units would report for common direction—that these decisions require a tie-in and that only by housing all the SHEP regulatory decision making in one organization can we expect to handle the important interactions in the regulation of technological activities.

Much of the indecision that interferes with innovation, slows up technological advance, and buys us nothing in additional protection comes about because of delays in the fragmented administration of the nation's regulations. If there were one unit from which all permits were issued, the whole process would be more efficient and take less time. The history of the last couple of decades is seemingly a story of projects endlessly delayed as they were pushed through hundreds of separate approvals from numerous independent agencies.

From time to time suggestions have been made by a few scientists that a "science court" be set up to settle the purely scientific aspects of SHEP issues. These suggestions have not been taken seriously because it is recognized by those with regulatory experience that in major controversies it is impossible to extract the scientific aspects and settle them in isolation. Moreover, the controversial issues tend to be precisely those where too little is known with certainty and where, consequently, it is difficult to find agreement by scientists on whether a given facet of the problem is a scientific one or actually a piece of a puzzling mixture of data and conjecture. Experience has shown that when the available facts will not settle an important issue and value judgments dominate, scientists can readily be found who are willing to put forth as authoritive their individual hunches about what dangers might be severe and what might be dismissed as trivial; the scientists are actually expressing their own value judgments.

The proposed board would from time to time certainly wish to bring together groups of scientists (from the government's own investigatory agency or from anywhere else) and let them debate issues—not under any sort of formal courtlike rules naively set up to control the discussion, but rather merely in an orderly attempt to bring out as many useful opinions and facts as possible for use in the board's decisions.

The board should be regarded as an extension of the executive branch of the government, that is, of the presidency—a conclusion we arrive at by focusing again on the board's main mission. We want the board to constitute a credible representation, a pragmatically effective microcosm, of the electorate. It should integrate the members' sense of values to form criteria for judging the options and then use these value judgments for decision making. The president should appoint the members with the consent of the Senate, naming citizens of outstanding competence and character, with staggered but substantial terms of office. This appointment process would cause the board to be inherently political in the way it should be: responsive to the country's goals and priorities.

In proposed technological activities, some private or public entity presumably always exists—a drug manufacturing firm or an electric utility—that wants to move forward with a questionable product or project. The entity represents special interests, or doesn't understand the project's negatives well enough to be deterred or believes those negatives are outweighed by the positives. In such a case, the entity and the proposed investigatory agency might at times appear to be opposing parties, one interested in the advantages of some activity it wished to start or continue, the other (the investigatory agency) ready to say what detriments it believed the proposed activity would cause. The pluses and minuses of the activity and the alternatives for regulating it would be argued as thoroughly as possible in the board's hearings. Perhaps these two interested parties would agree the activity was safe to carry on, or conversely, should be held up until identified disbenefits could be diminished. Whether they agreed or not, the board would settle the issue.

As the board went about its task of rendering regulatory decisions, the courts would sometimes be sought out by interested parties as in the past. If the legislation setting up the board were competently written, appeals to the courts might be expected to be less attractive and frequent than now. The board's decisions would be interfered with by the judicial branch only when the board overstepped its charter or ignored other pertinent legislation or failed to be just and fair. (For example, the board might make a decision it regarded as superior to any alternative for the good of the nation, but might overlook an injustice to some citizens in that decision.) No matter how we might improve governmental decision making as to technological applications, the decisions reached could end up in the courts. But we can do better than to encourage the presently growing trend of turning to litigation to settle most important issues.

Industry's Role

In an ideal world, business leadership would contribute greatly to sound regulation of technological operations. A corporation engaged in technological activities that might impair health, safety, or the environment would base its price structure in part on fully covering the expense of adequate investigations into possible disbenefits. Its equally ethical competitors would experience the same cost increments, so no disadvantage would be felt by a company properly sensitive to the potential negatives of what it was doing. All companies would respect the necessary mission of the government to establish standards and guidelines and to police to them. They would cooperate in making public all information unearthed in their own studies. They would propose modifications in the government's rules whenever they thought a good case existed for regulating somewhat differently—for instance, in relaxing standards. Whether or not their recommendations were accepted by the government, they would obey the government mandates meticulously.

This imagined characteristic of technological industry is unrealistic, and not primarily because it calls for idealistic management behavior. It is rather because of one inherent weakness in regulating to control the negatives of technological activities, namely, that when all is said and done, how safe we should try to be is still a value judgment. Perfect corporation managers could not insist that their products be perfectly safe in manufacture and use, since perfection is impossible. They know they must pit risks against prices, because effort to lower risks will increase costs. They would seek information helpful in arriving at sensible trade-offs, they would employ their own judgment in the risk-benefit performance of their activities, they would hope for market acceptance of these judgments where the market was the paramount decider, and they would rely on government regulations where regulations controlled.

But they would not set themselves up to be judges of exactly what level of benefits society should expect to receive from technology in view of the inevitable accompanying disbenefits. Thus, while a pharmaceutical company's leadership might come to possess considerable knowledge about the potential benefits and detriments of a new drug it had developed, elements of society well beyond that one company should participate in deciding whether the aid received by users of the drug should be deemed in balance with the danger from possible side effects.

Before legal requirements limited air pollution from automobile exhausts, what would have happened to an automobile company that of all manufacturers alone chose to add, say, $500 to selling prices so as to offer cars with negligible air pollution emissions? We know it would not have been able to market those cars. The motivation to a car buyer to make a tiny improvement voluntarily in the smog content of the air—with no one else required to do the same by law, and few likely to follow the lead—would be equally tiny. Price competition, often on a worldwide basis, would act to penalize that pioneer company trying to break through unselfishly to a higher standard. On the other hand, if a particular automobile were perceived by the buying public as substantially safer to drive than competitive cars, a somewhat higher price could be charged for it. Success in such a project would mean that the automobile manufacturer had correctly anticipated the value judgments of the customers, had guessed correctly how they would choose in the trade-off between diminished risk of injury and higher price.

In the real world, the best industry can do in working out risk-safety-price relationships is to anticipate and meet market requirements where they are applicable and government-imposed regulations where they exist.

Local Government and Regulation

When government was mentioned in this chapter, it was usually in reference to government at the national level. But states, counties, and cities also are involved in SHEP matters, sometimes supplementing and sometimes overlapping the federal government, and at other times engaged in independent regulation.[6]

Most of what has been said about the difficulties of and the possible improvements in government regulation of technological operations applies to all government levels. If the federal government improves the way it operates, this might set an example and some of the innovations might be adopted by other government entities.

---

[6] Only about a third of the regulatory programs of the EPA that the law says may be delegated to the states had been so decentralized when the present administration took office. The plan of the administration is to raise this fraction to 50 percent.

Local government groups should be heavily involved in minimizing the negatives of technological activities in their midst. The right trade-off between reducing hazards and increasing costs and prices often can best be realized if on-the-spot industry management, labor heads, and local government leadership all join in determining the balance. They should want to work out the regulations together because it is they who will live with the results. They should often be able to find balances and compromises more readily and wisely than OSHA can be expected to do for them. In order to lower a factory's accident rate, the production operation must usually become more costly. The product's price will rise and the probabilities of the company's success against its competition will be lowered. Pricing and company success considered against employee income and security is the key to labor-management negotiations. Management and workers have the greatest motivation to be both careful and innovative as they strive to balance safety against job security.

Consider, as an example, a small town with a chemical plant which is polluting the adjacent air, land, and river. The local government, representing the people of the area, presumably would press for the cleanest possible environment. On the other hand, the plant's managers might be able to show community leadership that to go further in clean-up activities would add to costs, increase product prices, and cause the company to lose business to competitors operating elsewhere. In turn, this would mean lower plant employment, which would be a hardship to the people of the area. A needed hospital wing, for instance, might have to be postponed because the tax increase to finance it would no longer be practical. Trade-off questions like these are always difficult to decide, but the local citizens are closer to the problem and can come to understand the facts and alternatives more easily. They, not the staff of an agency in Washington, will be the victims of poor decisions.

Regulation, Innovation, and the Triangle

In SHEP regulation of technological activities, the triangle of society-technology-liberty presents a clear picture. From the technology focus scientific advances emanate continually, promising benefits to society but also threatening harm. The liberty focus reaches out to press for freedom to innovate, to engage in technological activities without the encumbrances of too many restrictions. The society focus of the triangle insists there must be rules that will limit the freedom of individuals and companies or governments to do as they please with technology. The trick is to get the right balance—generous liberty and prolific innovations to spur the receipt of gains from technological advance, but appropriate rules to curb the harms.

The balances regarding automobiles are different from those of nuclear reactors or drugs or pesticides, yet they all interact. Successful SHEP regulation is an unending and constantly changing challenge to create and maintain balance. The most important thing to remember in meeting that challenge is that the three factors, society-technology-liberty, need to be understood as constantly influencing one another and irrevocably connected.

Chapter Five The New Power of Information Technology

The nuclear arms race is the prime example of the technological-social mismatch. Here social advance must come quickly or the technological progress may do us in. Information technology, in fortunate contrast, is developing just in time to rescue us from a problem. The information that has to be acquired and used to keep our complex society going is burgeoning, and the computer-communications revolution luckily is offering us means for more economical, reliable, and speedy handling of information precisely when we require it to avert chaos.

But this technological advance is not only the answer to a need. It also presents us with an opportunity for new heights of achievement. Human brain power is being extended by computers, electronic terminals, digital communication networks, microprocessors, electronic memories, and the rest. These electronic systems and boxes enable us all to be smarter at our jobs, obtain more facts quickly, and put information together at once to reach useful conclusions. This will make for higher productivity and lower costs to buck the inflation arising out of our failure to manage the nation's economy well. It will lead to more efficient use of resources and higher returns on investments and thus will stimulate free enterprise.

The United States today holds the world lead in most facets of information technology, although not all. We have the most broadly based research and development infrastructure to support the advance of information technology, and our large, singly integrated market for these products matches well to our technological position to net us an opportunity for overall qualitative and quantitative leadership in application of the new art. Information technology is also an area where the synergism between military and civilian technology is high, and American military effort in this field greatly exceeds similar effort by our international competitors. Information technology is the foundation for military command, control, communications, reconnaissance, and intelligence. The military has need for highly sophisticated technological implementations that often are beyond the immediate requirements of the civilian market but that later spur advantageous repercussions in commercial products. Being ahead of the Soviet Union in this technology means that we are better able to focus and direct our military forces, that we can make ours more effective than theirs using no more personnel or firepower.

Information technology is not like nuclear weaponry, which if employed can only set civilization back. It does not present us, as do some other technological advances, with the imperative need to minimize safety hazards as we generate benefits. The information technology revolution does confront us, however, with unprecedented opportunities which should be grasped and exploited, but which, if mishandled, could lead to harmful social trends. We could easily allow such negatives to develop, precisely because technology's applications are so advantageous as to be impatiently and carelessly sought.

Most of us—at work, when traveling, while being educated or treated medically, while being entertained or governed—are enmeshed with the information flow that makes these activities possible. In 1982, the number of electronic computers in the world, including everything from the big mainframes down to the microprocessors used in games, toys, autos, and appliances, exceeded 5 billion, more than the earth's total population. If information technology systems were installed today wherever an economic payoff could result, that is, in industry, banks, department stores, professional offices, airlines, hospitals, educational institutions, government buildings, and homes, then the immediate investment would add up to several trillion dollars. The size of such an investment and the returns on it, even when spread over a decade, obviously would have a powerful economic impact. The social impact would be even greater, because half the men and women in the nation could accomplish their tasks more efficiently by use of modern information technology, and this implies that those tasks would be handled much differently from the way they are today. Certain jobs would be eliminated and new ones created. In the process, while the general welfare of the nation would be enhanced, the patterns of many individual lives could be upset.

Rapid social adjustment never goes smoothly. The larger the economic gain to be realized by instituting change, the greater the pressure for speed in changeover. The higher that speed, the worse the dislocations. The more severe the dislocations, the more certain it is that political issues will be spawned. The government then will enter strongly, trying to manage the social alterations taking place. Government attempts to control things will create further confusion, and the government will find itself caught between those voters who want rapid change so as to realize benefits they see for themselves and those who oppose change because of the harms they fear.

The government cannot merely concentrate on directing or ameliorating domestic social adjustment, because the advance of information technology involves international parameters. As an example, the government must participate with other nations to divide up the shared and limited radio spectrum. It also must negotiate with other countries for the availability of positional slots in space for communications satellites. Since information technology is a powerful component of military strength, the government must be immersed in international relations on this front as well, not only in trying to outdo potential enemies but in assigning roles and missions in negotiations with allies. Thus procurement policies to provide NATO with command, control, and communications systems include doling out to the member nations the contracts to provide the equipment. The computer-communications industries within the NATO member nations compete with one another for the procurement contracts, even as their governments cooperate to form joint military plans.

Because information technology is fundamental both to economic strength and security, control of technology transfer from the United States to other nations by the U.S. government has become a major issue. Overall, the impact of the advance in information technology on international economic competitiveness is so great that the government is bound to be drawn into inevitable battles between nations for world markets in information technology systems and devices.

Information technology applications will lead to government-private sector interfaces well beyond the past experiences of telephone utility regulation and radio broadcast licensing. From the national level to townships, government must assign privileges for wiring up all regions of the country for the new video, computer data, facsimile, and other communications services. Government must referee the activities of the free-enterprise sector in information technology to avoid monopolies and other forms of intolerable unfairness, harmful influences on control of information flow in a democracy, chaotic financial consequences through new and unprepared-for liabilities, fraud when money moves electronically, and violations of privacy. Government's role is unusually difficult because, in addition to dealing with the many necessary nontechnological facets, government must also handle the technology itself which is especially complex. This requires that the government, as it attempts objective decisions on behalf of all the citizenry, possess a degree of technical expertise greater than is usually reasonable for it to encompass.

The free-enterprise system will contribute enormously in this field, but we can expect substantial instability for some time in the computer-communications industry. The more exciting a new field is as to potential growth and return on investment, the more readily available are investment capital and eager competitors. Races to dominate markets will be common. Some entrants will win, but more will lose out, and the competition will be fierce. (We recall the numerous early automobile companies and note how few names, many once proud, have survived.) Shortages of both technical and management personnel will develop. The sprouting of new, glamorous, highly successful industrial empires will be paralleled by bankruptcies.

Universities will suffer because of the rapidity of growth of the industry in this favored field of technology. As private entrepreneurs seek to realize their hopes quickly and beat their competition, they will make high salary offers to pull talented faculty members away from university posts. This short-range management action, already heavily at play, will ensure that within a very few years the flow of available graduates needed to continue the expansion of the information technology industry will decrease and become a bottleneck.

Finally, as we soon shall explore further, implementations of information technology can not only improve the efficiency of the handling of communications and running of operations, they can alter the way in which the citizens influence the workings of both government and the free market. Some added citizen participation in the management of the nation may be highly beneficial, embellishing both democracy and private enterprise, but other forms of citizen involvement made possible by the new technology may actually impede necessary social adjustment. Similarly, an all-out national embracing of computer-communications advances, while it would raise the overall capacity of the nation to obtain, process, and disseminate information, might also encourage the government to take extraordinarily tight control of the nation's affairs, a possibly somewhat less desirable development. Innovation soundly carried out in the organization and assignment of government-private sector roles is demanded by the advance of information technology.

The Expanding Role of Information

Information has not been made suddenly important by the advent of new information technology. Acquiring and utilizing information has always been fundamental to success in managing the nation's activities. The difference is that previously we lacked the means to obtain and use more than a small fraction of all potentially helpful information. Today's technology makes it possible to have much more information and do much more than the usual things with it. A new world of invention is opening up to reap rewards from the sensible employment of information.

An analogy will illustrate the point. Food, being essential, was sought and eaten long before humans learned to create and utilize fire. But after the technology of fire was developed, homo sapiens could boil, bake, roast, and fry. The new technology stimulated new food art, provided new enjoyments, and modified the culture. New tools and food processing methods were called for to achieve best results. Most important, however, is that fire and cooking technology greatly broadened the list of available, tasty, and digestible foods, removed poisons and bacteria, and made possible the masticating of matter previously not fit for human consumption.

Similarly, we now can have the privilege of procuring information and doing things with it which previously had never occurred to us. The definitions of what is useful information and what is cost-effective information handling have multiplied.

In every American's daily personal experiences, and in all business operations, ample evidence already exists of the growing complexity and necessity of information and of our dependence on information technology to make possible the handling of that information. Years ago, making airline reservations involved telephoning requests and keeping handwritten records. With the then relatively modest speeds of the planes and the number of flights and passengers, the information required to run the airlines and the means for processing that information were quite compatible. Today the airline counters all have electronic terminals, computers process the information, and signals flow automatically and almost instantaneously over national networks to identify passengers with flights and seats. Scheduling pilots, fuel, flight attendants, food, airplane maintenance, spare parts distribution, and all the other steps required to run a jet-based airline call for a tremendously greater information flow than in the past. Because of the increased traffic and speeds of travel, information now needs to be available quickly to employees distributed over thousands of miles. Without modern information systems it simply would not be practical to operate modern airlines, and without the new technology it would not be possible to obtain, process, store, and use the mandatory, rapidly changing information.

The credit card offers another excellent example of the expanding role of information technology. Granting a card in the first place begins with checking the credit rating of the individual, which means acquiring information efficiently from numerous organizations with whom the credit applicant deals, keeping the information up to date, withholding it from some and making it accessible to others. When a credit card is used, a stream of information is released that ends with the eventual transfer of funds to pay for the transaction. Handling today's millions of daily credit transactions would be impossible without recent advances in information technology. Information is the key to modern credit activity, and modern technology is the key to handling the information so that the system can work.

We are moving to electronic funds transfer, computer-controlled transactions that identify the parties to the transaction and items being purchased or payments being made; check credit or payment records; transfer funds from the buyer's to the seller's account; and provide information to the seller's system for inventory control and reordering. Money transactions between individuals, between individuals and companies, between companies, and between all of these and the government can be handled electronically. This means virtually instantaneous shifts of funds between payers and payees. New jobs are being created by the advent of electronic funds transfer, while jobs that previously involved handling money transaction information slowly, cumbersomely, and expensively are disappearing.

Industry's production operations are already using computers for scheduling materials purchase, fabrication, machine maintenance, parts distribution, inventory ordering, and the gathering and processing of data about the market in order to plan production volume. Earlier manufacturing automation advances were mechanical, involving the handling of physical materials to replace human labor or complement it. These advances achieved higher productivity and superior return on investment. The new information automation replaces or extends human brains in the control of the production process. This enables the production of much more per human hour expended and moves people away from manual effort to new positions where they direct, utilize, and are complemented by the electronic information system.

The president of General Motors recently stated that by the end of this decade 90 percent of GM's production will be under computer direction. Sophisticated robots with computers for brains, electronic devices for sensors, and automatically controlled mechanisms for fingers and arms will perform tasks in factories that are too dangerous, fast, boring, or hard on eyes, ears, lungs, and nerves to be well-suited to human operators. In addition, the jobs usually can be accomplished with less cost by using robots.

In the overall management of business and industry, companies soon will not be competitive if they do not make use of information that is probed, processed, presented on a broad scale, updated, and in control of the details of the operation. Included are these key business functions: managing cash and monitoring cash flow at all points of the system so that data can be constantly referred to computer-predicted models to anticipate and minimize cash requirements, thus lowering the cost of money and the cost of the final product; processing employee information to maintain personal records, ensure proper and legal treatment of employees, arrange government-required deductions, and schedule assigned tasks and career advancements; calculating, analyzing, and allotting funds for taxes; monitoring government regulations; and forecasting future markets, sales, earnings performance, and resource requirements. All of these functions depend on possessing information in increased amounts, more up-to-the-minute information, available almost instantaneously, that is, in real time, with higher accuracy, and with superior integration and interpretation as it is collected from widely spaced points. Handling information in this way is possible only because the advanced information technology is available.

In producing for the nation's needs, some of us spend our workdays doing things with information in constant interface with pieces of paper, telephones, input keyboards, and video screens. Others are busy using their hands or backs. For some years now in U.S. industry, the working population engaged in dealing with information has exceeded those who work with their muscles. Similarly, the missions assigned to government in our technological society increasingly require that more data come into the government, and there is greater need to process it for revenue collecting, economic planning, budgeting, regulating, and policing. Whether it be in industry or government, we have become an information society.

The Coincidence of Demand and Technology Breakthroughs

In the broadening application of information technology to society's needs, are we witnessing an accident, a lucky coincidence? It is surely remarkable that just when our society, with its growing complexity, interconnections, and size, requires more information and faster and more economical handling of it, science and technology have turned up appropriate breakthroughs. Or is it something else? Perhaps science and technology merely happened to arrive recently at the capability to offer us superior information handling, and then we adjusted the way we run society so as to make use of this new technology. If it is the latter, then, for those fearful that the society is becoming more technological than is good for it, the question becomes: Are we processing more and more information merely because we can?

Many executives, as soon as it becomes practical to learn more about what is going on in their operations, demand the added information. Most engineers, when they see that something very new and exciting can be done, want to do it. The military are always concerned that if we don't move quickly on everything we perceive is possible, a potential enemy may accomplish it first and put us at a disadvantage. Government bureaucracy wants to grow and delve more deeply into more aspects of society. For all these groups it might be that the more information they can latch onto, the more reasons they will invent for needing still more. To illustrate, we might invent a tale of a biologist of old who, upon the invention of the microscope, decided he needed one urgently so that he might take a visual voyage into the internal makeup of plant and animal matter, thereby coming to understand the many unknowns of which his life's work had made him aware. But when he latched on to his first microscope, he not only gleaned some long-awaited answers, he also uncovered numerous new questions he had not earlier known enough to ask. Despite his sudden progress, he found himself with a longer list of mysteries than he had before his powerful lenses became available.

This worry has to be reckoned as slight for the free-enterprise sector of the nation. Banks, airlines, department stores, and manufacturing companies will not suddenly invest in a lot of strange hardware—whose installation is not only costly but requires modifying the operation and familiarizing personnel with new practices—unless they have strong reason to believe that more abundant, better-processed information will pay off. It is one thing to risk buying a new factory machine, expecting it will perform some fabrication function more efficiently, and be ready to accept the loss if the results disappoint. It is another for a company to consider altering the way the activity is controlled. Here management typically feels that mistakes can be critical to the health of the firm and irrevocable.

But conspicuous coincidences do dominate the advent of the era of information technology advance. In the science and technology of computer communications, a number of diverse leaps have occurred simultaneously, and they have been remarkably mutually reinforcing. First, basic physics research has greatly increased understanding of electric phenomena in matter, right down to the level of the atoms and electrons that assemble themselves to form solid materials. This has led to new clarity on why some materials conduct electricity and others do not and, in particular, why some are semiconductors, capable of either supporting the flow of electricity or resisting it, depending upon subtle, controllable conditions. Assemblies of the right molecules (a transistor is an example) can be switched from one state (conducting) to the other (nonconducting) at high speed by imposing electrical forces. This has made it possible to put information quickly and cheaply into the form of a series of numerical digits (binary, not decimal, because in the binary system with only two digits, 0 and 1, the on or off electrical conditions can provide the immediate signal equivalent of either digit). All information—the words on this page, the balance sheet of a company, the census figures, a TV picture—can be put in the form of a series of numbers. The numbers can then be processed by semiconductor circuitry to create other information forms and serve up answers and conclusions inherent in properly manipulating the numbers.

Advances in solid-state physics, specifically in semiconductors, have simultaneously accelerated computer developments, because a computer consists in great part of an array of on-off switches. These can be interconnected and electrically energized to express the intelligence in a bit of information, sort information, perform logical operations (such as arithmetic) on it, store it and play it back, and do a vast number of other information processing tasks.

Microminiaturization of the physical apparatus to accomplish these tasks has meanwhile proceeded rapidly, first with thousands and soon after with millions of individual semiconductor elements fabricated automatically on single small chips. This has yielded enormous information handling power with amazingly small physical size, high speed, low cost, minute energy requirements, and high reliability. Information can now be expressed electronically and re-formed in ways that improve accuracy, make superior use of available bandwidths in the radio spectrum, and extend the transmission capacity of cables. Separate pieces of information can now be shaped and packaged together for lower cost transmission, then disassembled and restored at the receiver end. Communications between electronic information handling equipment and human beings has also improved radically because the powerful, although miniaturized, electronics circuitry can prepare the information for a remarkably useful variety of displays to the eyes.

Quite independently but coincidentally, it has become practical to locate substantial weights of complex equipment in orbit thousands of miles above the earth. There the apparatus can act as a radio relay, placing any two points on earth in direct communication. At about the same time we have learned how to put complex information patterns into the form of light waves, and how to form glass fibers of such internal makeup that they can transmit light waves containing far more simultaneous messages than copper cables at a small fraction of the latter's cost, weight, and size. (A glass thread thinner than a human hair, less than a thousandth of an inch in diameter, can carry the equivalent of a hundred thousand telephone conversations.) Finally, lasers have been invented that make it possible to generate unusually highly focused and powerful light beams. The combination of the various technology advances in the use of light rays make it possible to transmit the entire contents of the average public library in a small fraction of a second, a performance somewhat beyond the average demand.

What Information Technology Makes Possible

If we were to do everything conceivable with information that the new information technology makes possible, it would far transcend what it is sensible to implement for industry, government, the military, or the professions. It is not necessary or desirable to arrange that every person on earth be able to communicate virtually instantly with every other person, but the technological potential is there. If scientific and technological principles alone, not economic realities, were the limiting factors, all the information in the world's libraries could be put into electronic form and made accessible virtually instantly to anyone. Managers of companies could store every bit of information about the goings on in their operations and be able to tap the storage bank from their office desks, directing that any available information be displayed quickly for them to ponder; they could then perform numerous logical steps to derive conclusions for their further contemplation.

Information technology will never be used up to its technical limits because it would be economically absurd to do everything that can be done with information. Even if additional breakthroughs in information technology could make negligible the cost of any service humans could envisage, we would not always know what to do with all the information acquirable and would be inundated with it. Already the information that can be generated in many companies' computer rooms is far more than enough to occupy the full time of all the management personnel, even if they only glimpse at it as it flies by on their display terminals.

But information technology can now provide an enormous array of truly useful functions, and only a fraction of these tasks have yet been widely introduced. We have already referred to some. Here are a few others:

1. **Two-way wideband wiring of homes.** The resulting communications capability would cover not only conventional TV reception and telephony, but also would allow the home user to send signals requesting information from outside data banks. Equipment in the home would process received and stored information to aid the user in arriving at decisions. The home user could commit to purchases, pay bills, do banking, participate in educational programs, and set the system to monitor operations of household utilities and burglar and fire alarms.

2. **Reception of radio broadcasts originating anywhere.** Even locally broadcast programs in small population areas could be moved economically by satellite and cable into any TV set, even those thousands of miles away, yielding a much-expanded number of available radio or TV channels, beyond the present local stations and the few national broadcasting networks.

3. **Intelligent terminals in business and industry.** Almost all persons who today employ a telephone in their work could benefit in a cost-effective way by use of an appropriate computer terminal. Going beyond conventional telephone service, this would make possible the display and movement of a variety of information to and from that individual and other people, computers, and information files.

4. **Electronic mail.** Most of today's mail could be sent electronically from individual terminal to individual terminal—from person to

person or person to computer or computer to person. Each person would be able to read the message on the screen, store it, call it back as required, or order up a print.

5. **Teleconferencing.** Conferees located in widely separated regions could assemble for a video-communication interchange without leaving their home locations. They could see and speak with the other members of the group and display data to each other through use of video cameras and screens, voice-activated microphones, and long-distance wideband communication links. This operation would save time and funds, compared with the time and expense of conventional traveling to attend business meetings.

A Coming Robotic Society?

Because information handling has over the years become a high-cost portion of virtually all activities, and the new information technology has the potential to cut these costs, a rapid changeover to highly automated information flow would appear desirable to benefit the nation's economy. Moving swiftly to produce and install information technology devices and systems would increase employment generally, even as some jobs became outmoded, because the changeover, if done correctly, should produce an improved return on investment. This should make the economy grow and cover the costs of training people for new positions in the information technology age. In turn, this would make it easier politically for government to control inflation, limit government spending, lower taxes, and encourage investment.

But there is more to it than this.

Let us look more deeply at how information technology might affect the production of the nation's goods in the future. Superior management control over production surely will result from properly using all the information needed for close and speedy synchronization at each step of the process. The generation of power, the mining and processing of raw materials, the fabricating of all the bits and pieces, and the movement of the lines for product assembly are links in what ideally should be an integrated chain. The new technology can give each unit in the chain fuller and faster access to the information needed for solid control of its own operations. The supplier of materials and components can now be so well interconnected in critical information flow with the manufacturer-customer as to eliminate costly peaks, valleys, and bottlenecks caused by parts scarcity or excessive buildup and storage of inventory. Improved efficiency would result if the suppliers were to obtain their orders through electronic signals from their customers, who would in turn obtain their schedules in part from incoming orders that arrive automatically from their customers, and so on. Instead of the crude estimates that each corporation must now make of what operations will be expected of it each year, a supplier's operations could be intimately linked to the development of the receiver's demand. The whole would constitute a tight network of well-matched activity, eliminating waste, cutting inventories that result from wrong estimates of requirements, and saving time and money.

If such widespread industrial use of information technology were to come, it might spawn giant information networks that could store information as to what is being produced everywhere and could provide the pertinent information to management, shareholders, and government. Instantly, the flow of funds could follow the directed motion of materials and labor, as the information signals constantly changed to reflect the status of financial balances and to show who owned what and owed whom at any given moment (with the government automatically taking its cut in taxes). What we are describing is a system that eventually acquires and ties together economic control information not merely for a single company but for large collections of them. Of course, not all businesses need close ties with every other one—the printing of medical journals need not be scheduled in close coordination with airlines taking vacationers to Hawaii. But whether the integration is tight or loose, it could be ubiquitous in its availability.

When corporations agree to interconnect their information flow to the extent we have described, it amounts to new kinds of contracts and interdependencies. These must not be in conflict with the obligations of each corporation to its stockholders, creditors, and the government. For tax records and reports of corporate earnings, the new system's accounting mechanisms will be different from today's and will have to be arranged to be legally acceptable. Who is the properly stated owner of partially completed goods circulating between corporations in response to automatic electronic signals which neither alone controls and on which each has agreed to act? Who is responsible when something breaks down in the information system and losses in sales, profits, and investment result? The government will have to establish the rules to answer such questions. Indeed, it may even have to decide under what circumstances the interrelationships described should be allowed.

Today it is exceedingly difficult for the government to control the economy. Disagreement is widespread as to whether it should try to do so and to what degree. It is a subject of continuous debate as to how the government should attempt, through fiscal and monetary action, to influence money supply, interest rates, inflation, and economic growth. Even if all aspects of the economy could be monitored accurately, completely, and instantaneously, we lack adequate understanding of the key interrelationships and possess no credible, proven formulas to say what the government should do to achieve the economic goals it sets, assuming they are attainable in the first place.

Some years from now, however, when all the operating information of business and industry flows through interconnected systems, the situation might be entirely different. We might then have enough timely information and know-how to process and disseminate it so that business and industry can operate in close synchronization with a designated plan. Furthermore, the government might specifically set the plan, because it would be just a small step for the government to assume control over the information network and force the national economy overall to operate in response to government-originated signals.

Instead of waiting, as now, to see what economic negatives—inflation, unemployment, high interest rates—are building up and then moving to seek corrections by rather indirect macroeconomic policies, the government instead might set out to schedule in detail what is to happen and try to minimize the ups and downs. It could mandate schedules of production, employment, and distribution, all through the use of the electronic information system, its large computers reaching every nook and cranny of the economy to extract and deliver information. What would occur then would be according to schedule. All utilization of facilities and labor and the flow of matériel and dollars, so this scenario goes, would be maintained perfectly in step.

The ultimate condition we might reach, as the pervasive electronic information network is employed to coordinate American industry, would be one with the government in overall economic command of the nation. In that future technological society a mammoth government-controlled automaton would spell out for us where we are to work and live, what is to be produced and where, in what quantities and styles, where to go on vacation, whether to drink orange or grapefruit juice, and when to trade in our cars.

Perhaps the U.S.S.R. which has been trying to do exactly this for over half a century without complete success, has merely lacked the technology. They have sought central command of their economy, but have not had the necessary information system. Much more gathering, processing, and dissemination of control data would have been needed for them to accomplish what they intended. But perhaps the United States, with its advancing technology and with the natural trend that this technology makes possible, could actually make it happen some day in the future. If this unpleasant dream comes true, few of us would have to bother to think and decide. People would become anonymous cogs in a society of enmeshed and coupled signals, cables, gears, vehicles, flowing chemicals, and electric power, to the point that it would be difficult to distinguish the human component from the rest. The robotic society.

Is this the way it must be? Obviously, in such a world there would be little room for the citizen's free will. With firm, irrevocable connection between humans and machines under a rigidly communicated plan and ubiquitous information networks knowing and controlling all, how could we fit in individual incentive, pride in accomplishment, competition, initiative, and personal rewards?

Free Enterprise Made More Free

Let us look again at the technology on which this imaginary, extreme robotic society would be based. The dismal architecture of a government-controlled economy hinges on having the information needed for control and presenting that information virtually instantaneously to whoever or whatever needs it, everywhere, all the time, so that the whole system will work according to plan. But the moment technology arrives at the point where it can provide this kind of all-inclusive computer-communications system, we will also have the capability, for the first time in the history of humankind, to let everyone participate in deciding what happens—instead of telling everyone what must be done.

Any system for sensibly controlling the manufacture and distribution of the products we use in our daily lives will certainly be incomplete unless it includes consulting the consumer. Indeed, as the Soviet Union seems to have demonstrated, attempts at centralized economic planning fail if the citizenry is not included in the loop. The government's central computer can schedule the transportation and build the hotels and send forth the food so that people can embark on what the computer says should be their business trips and vacations. It can arrange the seating for particular airlines on particular dates and distribute the tickets. It can tell the public how many shoes, pickles, and bars of soap to acquire. The supersystem can arrange and watch the inventory of the resources of the nation and see that everything is neatly assigned and is distributed to its proper place. But if Americans have any opportunity whatever to buck the system and make independent choices, they will. They won't take the airline they were told to, or trade in the car as scheduled, or buy the shoes on hand. Even the Russians, docile as they are in response to governmental control of their lives, are known to defy somewhat the official manufacturing and distribution plans for consumer goods. They often refuse to buy what has been produced despite their severe consumer goods shortages.

Then why not start the scheduling by describing the alternatives and asking people what they prefer? Since resources are always limited and choices have to be made, why not ask people whether they want new cars or washer-dryers? A highly developed electronic information system surely is essential to a government-planned economy, but we can, if we choose, use such a system in the entirely opposite direction. The system could be employed to reach everyone with information about what is being offered and give people an opportunity to make their own selections.

The essence of free enterprise is a free market. All entrepreneurs with ideas and some capital to put at risk ideally should have access to ready means for putting forth their proposals and trying to influence other people, who have free choice, to buy those products or services in competition with others. In a completely free market, which we only partially enjoy today, all would be privileged to select what they want from the unrestricted possibilities presented in the market place.

Suppose the two-way TV set described earlier had been developed and installed in your home. Of course, commercials would still be offered, but instead of telling only about products already developed and available, they would equally often describe what manufacturers were planning to do, subject to your reaction. It would be suggested that you commit to purchase an offered product on the basis of the descriptions put forth, and there would be an incentive for you to do so. Automobile manufacturers would tell you about the models they were planning, and if you were willing to order well ahead of delivery time, they would grant you a substantial discount. You would buy by stepping up to your set and pushing the right buttons, identifying yourself, and creating a record of your commitment to purchase in the central information file. Having done this, you would expect that some of your fluid resources would leave your account and be moved to the account of the car manufacturer.

Sometimes you would first want to go look at samples on display. But during the course of a typical year, you might be willing to order in this way many of your purchases—food for the freezer, toiletries, home furnishings, clothing, automobiles, refrigerators, vacations, educational courses, and tickets for sports events.

With such a system, advanced information technology would fully provide all the beneficial aspects of planning. But now that planning would be part and parcel of free enterprise. The free market would be electronically enhanced. Implementation of manufacture, a detailed financial recording of all transactions, and final delivery would all result from the information processed out of the network. A chain of activities would commence with the offer on the free market and individual acceptance or rejection of it. Too few orders for a described product would cause plans for its production to be canceled. A large group of people committing to an automobile by means of the system would confirm the manufacturer's production plans, but would also generate information transfer leading to the right amounts of iron, coal, and chemicals to produce the steel, paint, rubber, glass, and plastic required, and would schedule the machines and the workers to make the parts and the assemblies with optimized timing. Thorough economic planning would characterize this system, but it would be in response to directives from the consumers rather than from a centralized government computer. This planned economy would be planned by the market. It could be accurately described as a planned *free* economy. It would retain all the features of optimized efficiency and cost reduction, yet be based on choices made by individuals.

Another immediate effect would be reduced risk for the entrepreneur and thus an enlarged opportunity for people with ideas. Today innovators have to commit capital well ahead of time, after relatively inadequate market research. They must gamble that the product will be acceptable at the price for which it can be manufactured and distributed, and risk the building of manufacturing capacity and inventory based on this guess. Tomorrow, fully assisted by information technology, entrepreneurs could put their proposals to the public beforehand for the cost of a TV query. Only if they found enough customers who would commit ahead of time would they go ahead, and they then could raise capital more easily and make plans with greater confidence.

Ideas beget ideas. Noting what was offered, others would come through with alternative proposals. Competition would stimulate innovation to combine what technology has to offer with what the people really need and want. Resources could be devoted more than ever to enhancing creativity and efficiency in every area: product conception, marketing, production, and distribution. This would result in a better risk-to-return ratio for the investor and would encourage investment.

Toward which of these two possibilities is our future society more likely to evolve? The government-controlled computerized information network that directs all economic activity? Or the use of the same technology to promote greater participation by all in a free-choice economy built from the consumer up? The answer hinges on our understanding that the new technology already is causing both alternatives to develope in embryonic form and that we still have the option to choose between them.

Instant Democracy

If we anticipate the gradual development of nationwide two-way information systems for consumer sales and production scheduling, the use of the same systems for the presentation of political issues and the gathering of the public's choices should be assumed as a natural parallel possibility. Of course, it would be nonsense to disseminate information instantly to all about every public question and ask for opinions in response. We could never countenance a democracy in which every time an issue surfaced, no matter how trivial and detailed, all citizens in the nation would be expected to vote on it from their homes. We shall always need to delegate, as now, the responsibility to representatives in government of doing most of this voting for us. If we fail to discipline ourselves as the technology makes new dimensions of information handling available, we could inadvertently bring into being an automatic law-changing and politician-firing device. Government would not work too well in practice if presidents, governors, mayors, or superintendents of schools had to receive confidence votes from the public several times every evening so they could attack their tasks with confidence the next morning.

Despite the vagueness of the specifics, a strong point stands out. In the exaggerated picture of a government-dominated society we presented earlier, individuals would act like robots and have less and less say. They would be told what to do by a computerized central authority that presumably knew everything and used this knowledge to work out what was best for them ahead of time. With no more exaggerated justification, we could postulate an essentially reversed application of the same technology in which, for the first time, the reactions and choices of the average person could become known immediately. This knowledge could be processed to provide directions *from* the citizens, not merely *to* them. A highly informed and participative citizenry could be one result. As far as the technology is concerned, this is just as easily implemented as a society of people who are dull, uninterested, and democratically defunct. The way the technology is used is up to us.

The Government's Role

We have pictured two extremes. Common sense tells us neither will represent the actuality. But we are certainly headed for a future technological society of wired cities and national information networks whose breadth, complexity, and versatility of information-handling services eventually will vastly overshadow the present telephone system. Financial transactions, production schedules, medical results, Beethoven symphonies, crime statistics, video conferencing signals, and all sorts of other data will move in and out of homes, offices, factories, professional offices, educational and hospital facilities, and government bureaus. Thousands of private firms will be involved in producing equipment and installing and maintaining the various systems, furnishing information, acting on the orders, and arranging for interconnections among the separate networks and with the users.

The networks will be an assembly of many millions of hardware items. With inadequate care as to the functions and specifications of the pieces and their interconnections, and as to the way in which information is formed and moved throughout the networks, the whole complex could develop into a chaotic jumble, a tower of Babel, unreliable, expensive, and only partially useful. While providing some advantages, it would come nowhere near fully realizing the potential benefits. It might neither protect privacy of information in situations that deserved it, nor make information available fairly, economically, and conveniently to people entitled to it. The numerous participants in operating the separately formed segments of the information system would each grind their independent axes. Monopolies might develop in critical places, and users would find themselves captives of those in control.

An independent referee is needed—a traffic cop, a rule setter, an objective applier of judgment as to the manner of operation deemed best for the nation. This function can only be performed by the government, granted that it can do so only imperfectly.

A technologically difficult but fundamental requirement is for the government to set standards. In what form should information be packaged, distributed, and presented so that it can be most efficiently stored, communicated, processed, and shifted from place to place, human to computer, printer to video screen, to satellites, to cables? We want to allow the maximum of freedom for users and designers to do with information what they choose. At the same time, we want to facilitate interconnections, and that means setting commonality rules for hardware and software. Moreover, we can avoid monopolies if we introduce and adhere to specified standards that make it possible for many suppliers to engage profitably in designing, producing, and installing apparatus and storing and offering information.

Telephony and television became economically sound only when large-scale production and use became practical. Similarly, most of the new functions that information technology now is making available will be cheaper, more reliable, and generally more beneficial if mass repetitive use is made of the hardware and software. This also requires considerable standardization. However, it would be wrong, both economically and creatively, to erect severe, rigid restrictions on what can and cannot be done. Overcontrol by government would restrict progress.

Some coordination towards sound compromises will occur naturally through the free market. In the past, with TV standards as an example, participating companies have been known to get together and agree upon common approaches in order that all might share in what then could become a bigger market. But full use of information technology on behalf of the whole nation involves so many more dimensions and so much more complexity that similarities with radio or TV or telephony are highly insufficient as a guide. The desire to accelerate the availability of the benefits of the system, while also arranging for the maximum of competition and the broadest range of input and creativity from all potential suppliers, dictates an essential role for government.

Setting appropriate specifications for the handling of information requires understanding in detail some of the most technically difficult concepts in all of science and technology. The most complex calculations skilled astronomers perform require no more unusual a mathematical mind that the sophisticated logic needed for computer-communications systems architecture. The manipulation of information in electronic form, the automatic separating and packaging of bits of information, and the preservation, reassembly, and extraction of the original intelligence contained therein—after it has passed through numerous stages during which interconnectability of equipment is fundamental—are only a few of the facets included in the task of designing information hardware and software. To handle such efforts requires individuals of unusual talent, individuals with long periods of education and extensive experience. Naturally, it is very difficult for the government to attract and hold even the minimum of expertise needed to do its part of the task. Accordingly, the technological industry and the universities, where the capability is mainly to be found, must assist the government. The government needs to be the chairman and to retain the responsibility, but it must call on competence from the private sector or its mission will not be performed adequately. If the government falls down on its job, then the private sector will not be able to do its work well, either. The nation then will fail to make the best use of advancing information technology.

Much of the implementation of information technology requires the cabling of the nation. We already did this once over the last century with the telephone system. Now we are engaged in doing it again with video cable systems. In both instances, the sensible plan is to allocate the privilege of providing wiring through geographical monopolies granted by government to individual companies, and through this practice avoid costly and confusing overlaps and duplications.

Arrangements made so far are only a fraction of all that will happen as information technology blossoms. Industry, homes, offices, hospitals, schools, and government units will generally profit from being cabled so that audio messages (conventional telephony), TV programs, and all forms of data for computer or human utilization will be received and transmitted in both directions. For many years the existing telephone network will remain the most extensive existing portion, augmented by the separate one-way cables for entertainment TV reception that already exist in many areas. But eventually the problem of doing it right will have to be faced. The ultimate network should be one that will reach almost everyone everywhere with a single system of wideband, high information capacity, two-way cables connectable to every form of input and output device, from human beings to electronic and mechanical apparatus. In the interim, as we progress slowly toward that optimum installation, we are bound to live with a hodgepodge of partial installations. Even though the government may try very hard to prevent it, substantial disarray will result from systems that do not interconnect.

A National Information Utility

We might imagine that in the future one huge corporation, an information utility, would emerge to install and own a single supersystem operating throughout the nation. Of course, such a utility would have to be regulated by the government as to rates and quality of service. The system might consist of all the apparatus and interconnections required to receive, transmit, and deposit information. Numerous private suppliers might compete to furnish the hardware and the operating software to that information utility. Other organizations, operating on a free-enterprise basis, might be in the business of acquiring useful information, storing it, and selling it to those who request and pay for it through the system.

What we have been describing, though a far larger task, is akin to what has been the AT&T's telephone system (although AT&T covers only 85 percent of the nation, the rest being served by a large number of separate smaller utilities handling local telephonic communications and interconnecting into the larger AT&T long-lines system for long-distance telephony). However, the scale of activities is much greater as we move from narrow voice-to-voice telephony to the full range of information handling. For the broader service it is not always easy to separate the moving of information from the processing, storing, packaging, repackaging, and originating of it. Transmitting information economically oftentimes requires doing a great deal to reshape it so that the communication costs will be low. Separate information chunks must be made to share the costs of expensive transmission links, minimize idle time on elaborate apparatus, and make use of the various channels of communication. These may range from wires and cables to radio links, satellites, microwave antennas on mountain peaks, and then back into wires and cables.

A "chicken-and-egg" problem exists in design of the system. The pieces cannot be specified until the whole system is conceived and understood. In turn, the system cannot be designed properly until it is clear what kind of components will be available and what their detailed performance specifications will be. Those companies manufacturing computers and all of their ancillary equipment, such as input-output devices and printers, very often need to create ensembles of their equipment to serve a customer who has large-scale data-handling problems and a need for interconnection between installations spaced from a few hundred feet to thousands of miles. It has become almost impossible to compartmentalize such computer installations and design them in isolation from the communications system that will interconnect them with each other and the humans involved. Similarly, modern communication of information involves the use of computers to switch and alter the form of the information for practical acquisition, storage, movement, and presentation.

Thus a company previously envisaged as in the communications (or even more narrowly, the telephone) business, such as AT&T, now has to be seen as also in the computer business. Such a company, strong in systems engineering, often will be the best source for various other functions of handling information, including designing and providing user terminals well beyond the ordinary telephone handset. Equally not surprising is that companies previously envisaged as computer companies (such as IBM) find themselves also simultaneously in the communications business. Information technology now can be pictured quite accurately by pairing the two words, computers and communications. As a final nonsurprise, AT&T and IBM, both using technology to do what comes naturally to serve their customers, now find themselves competitors, each covering a broad range of computer-communications services. This same new competitive consideration applies to a host of smaller companies previously classified separately either as computer or communications companies but now better called computer-communications companies.[1]

Meanwhile, for many years the Justice Department has been suing both AT&T and IBM as monopolies in their two supposedly different fields, AT&T in communications and IBM in computers. Congress and certain other parts of government, such as the Department of Defense, the Department of Commerce, and the Federal Communications Commission, have recognized the inevitable (in fact, the already existent) merging of communications and computers and have been plotting how to release AT&T from prohibitions against the natural expansion of its communications role into computer-communications. Thus the government has realized AT&T and IBM are competitors—while simultaneously seeking to break both of them up, claiming each is a monopoly.

After a decade of pursuing monopoly ghosts, moving very expensively in the wrong direction, the government recently dropped the antitrust case against IBM. Almost simultaneously, it arranged the splitting up of AT&T. AT&T will separate from its thirty or so operating telephone companies handling local communications across the nation, freeing the parent company to provide equipment of all kinds to everyone competitively. Moreover, AT&T will offer long-distance interconnections to its divested operating units and to all others who desire long-distance service. AT&T has been granted no monopoly in any area but will rather compete against others offering alternative means for long-distance information transmission.

Accordingly, the idea of a future supersystem company providing national information services as a controlled monopoly now appears academic. Since the government has chosen to remove this idea as a possibility for narrow telephone service, it certainly would be unwilling to allow it for any broader range of information handling. In ruling out large information utilities, the government has made its own problem worse. Let us consider a

---

[1] New combination words like *"compunications," "compucations," "commputers,"* and *"commuters"* are regularly invented, but none has caught on so far.

number of specific examples where government refereeing is necessary and observe some of these difficulties.

Electronic Banking

We noted earlier that we already have become a credit-card society. A large and increasing fraction of consumer or retail transactions is arranged through prior credit granting plus quick credit checks at the time and point of sale, accomplished by inserting a credit card into a machine and pushing buttons to initiate electrical signals that approve and consummate the transaction and record it. This concept of electronic funds transfer (which has spawned such terms as "electronic money" or "cashless society") has been with us in embryonic form for a decade or more. It is rapidly becoming a widespread reality. What is involved is more than the originating of electronic signals at the moment of sale (in place of an exchange of currency) or the writing and processing of tens of billions of bank checks annually. Electronic funds transfer is creating a new dimension in banking.

When payment is made through electronic signals, then we do not need conventional bank accounts. The transaction entity handling the electronic signals (an information utility or credit-card granter) has a data repository that is instantaneously updated when the transaction occurs. The system can make an interest charge if the balance of funds happens to be negative for a period or pay interest if it is positive. If we were to use such a system for all our purchases or payments, no need would exist for paychecks to be deposited in checking accounts. We could simply arrange for our employers to deposit our income (electronically) to our account with the transaction system. The same arrangement could be made with corporations that pay us dividends. Credit card granters (Visa, MasterCard, American Express, Diner's Club, department stores, oil companies) are, in effect, in the banking business. Or are they? How do we define banking? Is any company to be allowed to handle money for others? Who is to be permitted to gain financially from the handling of funds as they move electronically? Who is to decide what the interest charges shall be? These are questions that presumably should be handled by the federal and state agencies responsible for regulating banking.

The maturing of electronic information handling will surely affect existing commerical banks greatly. For one thing, they will have many new competitors for some of their important money-holding and -moving services. If funds move instantaneously through electronics, then there is no float and this changes the effectiveness of the use of cash assets. Automatic terminals now make it possible for bank customers to withdraw cash at any time of night or day if only they insert the proper card and push the right buttons on electronic machines. But much more can be done by electronics. As electronic signals replace checks, withdrawal slips, deposit forms, mortgage or loan payment receipts, and such, banking customers need never show up at the bank for mere transactions. They can identify themselves, insert instructions, and accomplish their purposes if provided with the right kind of keyboard and card entry device in any shop or building, at home, or on a street corner—in short, anywhere the appropriate terminal can be placed. The terminal can be connected electronically into the information system hundreds or thousands of miles distant from the location where the bank supposedly is authorized to do its business.

Today's state banking regulations that limit what banks are able to do for clients outside of their home states are seen to be inconsistent with the potential of the new electronics. Also, they are not in tune with what other entities in the nation that issue credit cards already are permitted to do for their clients across the entire nation. Electronic information handling will force the government to rethink banking and set new rules. Failure to do this in a timely fashion will inhibit the availability of economic and convenience benefits which would otherwise be permitted if the new technology is properly applied to money motion and storage.

Privacy and Security in the Information Technology Society

Privacy and security take on new meanings, present new difficulties, and create requirements for government involvement when most of the information controlling society is handled by automated computerized information systems. In some ways, no fundamental difference exists between a manila folder filled with longhand notes and an electronic memory when each holds information about a person's credit rating or net worth. In both instances, if dishonest people get at the files, they could commit fraud or do damage to individuals whose privacy is thus invaded. The detailed mechanics of improper access to the two files are very different, however, and hence the ways in which it is possible to prevent access also vary greatly.

It is practical with the new information technology to keep track of information about people to a depth beyond anything we could have attempted before, and to control money flow with less participation by human operators. Technically the system makes it possible for the stored information to be deposited, moved, updated, and made available far and wide and much more quickly. Those who design and operate the systems must have awareness of this and anticipate the possibilities of invasion of privacy and crime when they design the system's characteristics. To be certain that proper precautions are taken in design and operation, the government should set some minimum standards for privacy and security performance. It has been traditional for the government to guard the depositors of banks against carelessness by the bank in the handling of funds. The same must apply in the electronic information age, and the government needs to spell out what all storers and handlers of data that control money must do.

Controlling access to civilian information systems bears some relationship to the military's problem of policing the availability of secret data. The military naturally is very advanced in the use of electronics for coding information and creating electronic equivalents of keys to ensure access is available only to those authorized for it. The military will be reluctant to disseminate its encryption technology for commercial applications and, in fact, will insist on protecting its coding tricks. Meanwhile, civilian information systems will grow in importance and will be increasingly designed so as to protect privacy and security. The very encryption techniques the military would like to keep secret will be reinvented and applied to the civilian applications. Cryptology is becoming a civilian field of engineering, even as the military tries to classify each coding design that surfaces, seeking to prevent its public exposure. This is another dilemma for government refereeing.

Let us consider another inevitable government function. In time, both private sector and government units will build up numerous useful electronic data banks that many will want to access. The government information inventory (such as the Labor Department's figures on production and employment, the Census Bureau's statistics, or the data of the departments of Commerce and State on domestic operations and international trade) will be of increasing value to industry and other nongovernment entities. Meanwhile, seeing that industry and the public at large are finding it valuable to have quick electronic access to all kinds of information, numerous entrepreneurs will arise to be collectors and maintainers of a variety of information banks. They will charge customers for access to their files through the national information networks. Again a problem will arise as to the privilege of selling information through the networks.

The government will need to set quality standards for information to be delivered to the public over networks that generally will be operating under government-granted licences. This might include limiting badly prepared information, misrepresentation, pornography, and harmful propaganda (for instance, hate literature). Doubtless, free market considerations alone would cause some would-be purveyors of information to succeed while others fail. However, the censorship issue may be perceived as different for a network that reaches all homes than it is for print media products sold at street stands. Whether there really is a difference is partly what has to be decided.

Electronic Mail

It is possible today to move information between individuals or entities in the United States by using the postal service run by the federal government. There are numerous competitive ways, from telephone, telegraph, and facsimile services to private mail delivery, to send messages. But the Postal Service provides a basic service which is not likely to be abolished. As new electronic techniques expand, conventional mail service may suffer from the competition. In principle, the Postal Service has the choice of sticking to its existing letter delivery system, leaving its service unchanged as technology advances, or incorporating new message-delivery approaches based on advanced technology.

So-called electronic mail is already being espoused and offered in some regions by the Postal Service. The sender originates the message in specified electronic form by typing it into an appropriate terminal (located in the office or home or local post office branch) that automatically transmits it by telephone line to the Postal Service's terminal nearest to the addressee. Here it is automatically converted into a printed letter, placed in an envelope, and delivered by the local mail carrier. Numerous variations of this approach can be imagined. In the electronic information age, how should we now interpret the government-established monopoly and protection the Postal Service has enjoyed? What is the proper role of the Postal Service? How should the laws be altered so that the Postal Service can serve the nation best as technology advances?

The Postal Service has already announced its belief that it has a monopoly on first-class mail, that this monopoly includes electronic versions of first-class mail service, and that it would like to use radio broadcasting for some aspects of electronic mail delivery. This has been criticized as putting the government improperly into competition with private movers of electronically expressed information. The claimed monopoly privilege of the

Postal Service also has been disputed by the Federal Communications Commission, which guards the available radio spectrum and objects to allowing the Postal Service access to the radio waves. The spectrum, according to the FCC, should be preserved for licensed, competitive, private participants; it is unfair for the government to compete for the limited bands. This disagreement is best regarded as unfinished business. It will have to be attended to very soon if the electronic mail idea is not to be hindered.

New Policies for Radio Spectrum Allocations

The government, through the FCC, has always had the mission of assigning privileges for radio broadcasting. In exercising this responsibility, the government has had to ensure that the frequency spectrum is divided efficiently and to arrange that broadcasters' signals do not overlap and interfere with each other. The new advances in communications technology that have brought satellites and cable systems strongly into the act have made it possible for programs from previously local broadcasts to be made available to points very distant, without interfering with the local broadcasts in that distant area. To the citizenry this offers a large number of channels and makes it economically feasible for programs of a more specialized nature to be available everywhere. (By contrast, consider the present situation in which, unless a program is attractive to several tens of millions of people, it loses out to the more popular programs that bring greater economic returns and thus preempt the highly limited broadcast channels.)

The advent of greatly increased channel availability changes an important aspect of the FCC's duties. Rather than having to control quality of programming in the public interest, the FCC now can sensibly rely more on the public to determine what programs are presented. When only a few TV programs are offered for selection, broadcasters given the privilege to monopolize a channel should be required to meet prescribed content standards. However, if the public is able to choose from fifty to a hundred programs, then none can hope to possess the viewers' attention by default. To be successful against that much competition, an ample number of channels surely will meet the public's idea of quality because those that do not will lose their audiences.

The new technology thus offers relief to the FCC in some aspects of its task. But it makes others more difficult. For instance, communications satellites generally operate in what is called a geostationary orbit. (When placed in an equatorial orbit some 22,000 miles from earth, the rotational period of the satellite about the earth is twenty-four hours and it has an unchanging position relative to the earth below.) While satellites and their protruding antennas are not likely to be over 100 feet in their dimensions, the practicalities of allowing many satellites to be in a synchronous orbit simultaneously requires substantial spacing between them. Moreover, the satellites' signals must not interfere with each other and with other transmitter-receivers, so they must have allocated frequencies for receiving and transmitting back the information they handle.

Economic considerations require that communications satellites carry many channels of information rather than a few. Furthermore, the ground pattern of effectiveness—that is, the directivity, or focus, of the beamed radiation in and out of the satellite—must be concentrated on specific geographical areas on the earth below. The simultaneous allocation of space position, frequency band, and radiation pattern constitutes a broadening of the FCC role (as compared with its role in conventional radio and TV broadcasting). Moreover, this allocation must be done on an international basis. All nations of the world would like room in space reserved for their satellites, not only to cover their domestic communication interests but to permit them appropriate involvement in international handling of information (whether it be for telephones, TV, computer-to-computer data, electronic mail, or any other application). Clearly the U.S. government must deal with other countries on these matters on behalf of both the nation and the private entities who are participants in information-handling systems that include satellites.

Easy Technological Innovation—Difficult Government Regulation

Technological innovation in information handling is not a job for government. The universities do the basic research, and the technological industry is responsible for development and implementation. The economic rewards for sound innovation are great, so the motivation for free-enterprise investment is strong. Technological innovation thus will flourish. Technological innovation, however, must live side by side with, and be supported by, correct government rule making if there is to be effective implementation of the potentials of advanced information technology in America. If the government performs well, providing integrated leadership for those functions it alone must handle, then the private sector will do a superior job of technological advance. Those contemplating risk investments will see a rational, stable, and continuing environment for long-range development of information technology. The free market then will work to optimize investments.

The interface of government with information technological advance is multidimensional and so intimately involved with private-sector decisions that surely the relationship of government to private endeavors is both the key and the limit to progress. In defining and implementing the government's role, there are some fundamental problems. Here are examples:

1. The government's responsibilities are loaded with technological-social-economic-political interactions requiring a systems approach for effective execution. That is, all the factors must be dealt with in an integrated way. Yet the government is a fragmented organization, and its many segments operate in isolated, piecemeal fashion.
2. Long-range views and policies are essential. Yet many of the government's units that must be in the act are dominated, out of practical political necessity, by short-range pressures.
3. The government's task is enormously complex. Yet the government, being a large bureaucracy, is not suited to handling complex problems. These problems can better be solved by small groups of outstanding individuals with authority commensurate with their responsibilities.
4. The government's actions in bringing the benefits of advances in information technology to the nation should preferably be straight-forward, with direct control and clear responsibilities concerning the impact of these advances on the private sector. Yet government in America tends to be equivocal and unclear, with overlapping responsibilities and much of its influence indirect and wavering.
5. The basic technology is new and difficult to comprehend, yet it is unlikely the government can attract and retain enough outstanding experts in the pertinent technology.

In view of these considerations, and recognizing the breadth of the government's mandatory responsibilities, one guide for assigning appropriate roles to government and to the private sector is that the government should not be asked to do more than is necessary. For example, the Postal Service should not expand to embrace more functions in information transfer than its present duties plus their most natural alterations and expansions in electronic directions. The government should not create an agency to develop and operate a national network to store and move information about, or to process credit information for a checkless society, or to provide information on request to homes, offices, and factories through a huge government-controlled electronic library.

Needed: A Broader Government Agency

Congress has been working for years to redo the basic legislation that affects information technology, a badly out-of-date (1934) communications law. No center of responsibility and policy integration exists in the government in computer-communications. To fill this critical void, Congress has considered creating a new national information systems agency to take on the control of all aspects of the government's duties in matters of information technology. If taken to an extreme, this would not be a practical approach because most information technology issues also include issues outside the area of information technology per se, considerations requiring a different kind of coordination. For instance, the Federal Reserve Bank certainly has to be foremost in authority in electronic funds transfer and its impact on banking. Similarly, the Defense Department cannot let go of prime responsibility for information technology matters that relate to national security. Again, controlling information technology transfer to other countries must involve the Commerce and State departments as well as Defense.

However, the fact that a single agency cannot house everything the government has to deal with in matters related to information technology does not require that we settle for the present disconnected, widely dispersed responsibilities and fail to cover in a timely way the new, essential tasks that are surfacing. One government unit within the executive branch could be assigned much, though not all, of the leadership. That organization also could be chartered to study and investigate the entire range of information technology functions of the government, even where it cannot be given the action authority. It could be charged to compare alternatives, always in light of the combination of technological, economic, social, and political facets involved, both domestically and internationally. It could act as an integrator to bring in expertise from technological industry and academe to supplement its own in-house staff. This agency would then be the chief assembler and recommender of policies and actions for all governmental activity involving information technology.

But the agency need not be limited to the investigatory and advisory role. Control, allocation, licensing, and regulatory aspects now in other agencies should be moved to it to the maximum practical extent. For example, the new agency could take over the broadcast licensing duties of the FCC. Better yet, the new agency could be the FCC, expanded to include the other responsibilities already described. The broadened agency would not usurp the task of controlling banks as their functions become increasingly electronic because there are too many other aspects of banking that have nothing to do with electronics. On the other hand, the agency could materially affect the needed legislation and the modifying of policies and activities of the banking regulators and remove from those regulators the burden of having to understand and cope fully with the complexities of the information technology revolution. The banking regulators, as they adjust to the age of electronic funds transfer and the development of national electronic banking networks, could lean heavily on the expanded FCC.

Similarly, the enlarged agency would not control information technology advance in the military field. The Department of Defense would continue to do that, of course, but the DOD would now have a place to go where integrated studies were made of how the advance in information technology was affecting civilian activities of the nation. This would militate towards the fullest exploitation in the civilian sector of military technological advance, and it also would aid the military in certain of its priority problems of command, communications, and control. Consider only, for example, that the warning networks of the nation vis-a-vis potential strategic weapons strikes against us, as well as the command system used by the U.S. military in the continental United States and NATO areas, include as key elements some of the existing civilian communications networks.

The FCC would need to add a "C," be renamed the Federal Communications and Computers Commission (FCCC), and be broadened in scope. Then it would be able to handle the government's part in making the triangle of society-technology-liberty work in the information technology arena.

Chapter Six The Dimensions of Space

We have discussed so far three of the very different faces science and technology present to society. The first, nuclear weaponry, illustrates that some technology will serve society best if never employed. The second, government regulation of technological implementations, discloses the importance of achieving a sensible balance between the benefits and hazards of technology. The third, the computer-communications revolution, shows that technological advance can yield bountiful gains if we organize to reap them.

Space—its exploration and utilization—is a whole other thing. Space (and here a cheap figure of speech, in most ways unforgivable, is peculiarly apt) is out of this world. This is so whether the project involves landing humans on the moon, linking the major continents with satellites for live TV broadcasting, or sending a spacecraft to the outer planets of the solar system. A quarter of a century ago space emerged suddenly as the arena for a science Olympics between the United States and the Soviet Union. Now space is a potential region for warfare and an indispensable tool for the verification of disarmament and, hence, the prevention of war. Space is a multibillion-dollar civilian growth industry and an inexhaustible frontier for scientific research.

The world is now putting hundreds of thousands of pounds of equipment into space annually. The government-private sector relationship for advancing space utilization is setting precedents. Finally, more than any other field of scientific research and technological applications, space is international, for it is a three-dimensional ocean that surrounds all nations and that they are forced to share.

How did it happen that the U.S. space program started when it did? And how did it immediately rate a high priority, a several-billion-dollar annual budget, a brand new agency under the president to manage it, committees devoted to it in both houses of Congress, and more attention by the world's communications media than any other science or technology program in history?

There's no mystery about it. The launching of the Soviet Sputnik in October 1957 surprised the world, but it shocked the United States. We had thought that while the Russians excelled in some narrow specialties—ballet and caviar were then considered suitable examples to cite—when it came time to launch an artificial moon, we would be the ones to do it. In fact, at that moment we were readying a modest satellite project, placing in orbit a very small instrument package, to be part of the publicly announced International Geophysical Year. When the Soviet Union upstaged us, it was insulting and alarming. If they could do this, the convenient inferior status we had assigned them in science and technology could no longer be assumed as realistic and they might outdo us as well in military weapons systems. In addition, the Russians, it now became clear, had been deceiving us, keeping from us the fact that they were not technologically backward. Our distrust of them suddenly became boundless. We reacted emotionally to the fear and the dare and sought to wipe out the humiliation of being bested so conspicuously in technology, a field of contest in which we had rated ourselves the acknowledged world champion.

The Space Race

Thus began the space race, which led to our inventing the major event in the new Olympics, a manned lunar landing. This world spectacle, the boldest space feat then imaginable, became symbol and substance for our regaining the lead, not only in space, but in all science and technology. After the Soviet Sputnik blow, the president created the post of science adviser to the president. Interest in higher education in science and engineering ballooned, and the government increased all its R&D budgets. Space technology was not only launched as a new priority category; it became also the spearhead for accelerated efforts on every scientific and technological front.

The U.S. entry into the space age was an accident in timing. It was not that, having looked carefully at all that science and technology might next make possible, and noting our most urgent needs and exciting opportunities, we decided the moment had come to commence space conquest with all rocket nozzles aglow. Nor did we opt for space in the late fifties merely because we happened to have arrived then at the point in technological advance where we could. It is true that once the grand-scale program to provide an intercontinental ballistic missile (ICBM) system became mature, we enjoyed as well the ability to orbit a substantial payload. The American ICBM program, begun some four years earlier than the Sputnik launch, had developed the entire range of technology: large rockets and matching fuels, light yet strong structures, electronics for control and guidance, reentry techniques, production lines turning out reliable quantities of hardware components, test instrumentation, and large-scale launching and tracking facilities stretching out over the Atlantic from Florida. The earth being round, an ICBM capable of delivering a substantial weight accurately on a target half of the earth's circumference away could—with a somewhat smaller pay-load—be made even more easily to overshoot and miss the earth entirely, thus going into orbit around it. Moreover, a mere 50 percent increase in the dimensions of the ICBM rockets corresponded to a tripling of the payload capacity. The ICBM technology accordingly could be extended to place human passengers in orbit and provide them with oxygen, food, reasonable comforts, suitable communication with the earth, and protection during reentry.

This meant that by the late 1950s, we could commence satellite and other manned and unmanned spacecraft projects. However, if the Russians had held off on their Sputnik for years, we would have done so with our space efforts as well. The U.S. space program would have been started later and would have been less frantic. The first major goal might have turned out to be a spy satellite or a commercial project with immediate and high return on investment, such as an intercontinental TV or telephonic link by satellite. It would not have been to put a man on the moon. Whatever might have been seen as the rationale for injecting people into near orbit or farther out, such adventures would not have needed rushing. The curiosity of scientists about how biological matter might react to a gravity-free environment would have stood in an orderly line along with their inquisitiveness about numerous other frontiers of knowledge, all to be addressed in due time. Indeed, President Kennedy's announcement of the goal of a lunar landing and return before the end of the decade of the 1960s came as a surprise to the leaders of the scientific establishment who had thought then that they had already put that proposal in its proper place, assessing it as less deserving of support than many others.

Those who had argued that the enormous funds for the manned lunar landing project (around a hundred billion 1983 dollars) should be spent on other things—creating more Harvards and Caltechs, broadening astrophysics research by observations from unmanned spacecraft, enhancing microbiology, advancing high-energy particle physics to explore more deeply the nucleus of the atom, seeking a cancer cure—underestimated the backing for the contest. If we had wanted only to check out the moon scientifically, we surely could have done so more quickly and cheaply by sending instrument packages there, including even a device to pick up moon rocks, take off again, and return them to earth automatically. But putting humans into a spacecraft and heading them for the moon satisfied our psychological need as no other competitive project could. The successful manned lunar landing replaced American feelings of newly found inferiority with newly confirmed superiority. Concern turned into exhilaration. As a momentous achievement, visiting the moon pushed all else temporarily into the background. American astronauts placed on the moon seemed to reestablish all Americans as leaders and pioneers and delivered us again into a world of promise and discovery we could all share.

The Military Role of Space

After the Sputnik surprise, the American public quickly associated the demonstrated technological prowess of the Soviet Union with a military threat. This was a vague feeling, but it was strong and had to be attended to. Dwight Eisenhower, not only the president but perhaps the most highly regarded military professional and father figure of that time, said there was no reason for worry on the security front. But that reassurance was not sufficient. We needed a large-scale space effort, a view particularly forwarded by two groups. One, a growing fraternity of rocket enthusiasts, had been waiting patiently and not too optimistically for many years to journey to the moon and the other planets. To them, the Sputnik success and the public's reaction to it were very good news indeed.

The others were the zealots for military preparedness. Going much further than the average citizen's insistence on adequate military strength, this group had a simple rule to relate scientific and technological advance to national security. The rule was that since all advances might lead to military advantage, it was vital that we be the first to develop every potential or we would lose the next war. These ardent supporters quickly called attention to the possibility of space warfare in which spacecraft would attack other spacecraft and each side would seek numerical and firepower superiority in space, intent on full exploitation of space while denying it to the enemy. The moon, they announced, must be captured at once and turned into a solely American platform to be used to bomb the earth or as an invulnerable hiding place for nuclear weapons. They warned, "The moon is the high ground, so whoever controls the moon controls the earth. If we don't move fast, what will we find when we land on the moon? Russians."

Twenty-five years later, space satellites have indeed been found essential for certain military functions, but the moon, despite its best efforts to volunteer, has not found a war role. This natural satellite of the earth (which turns out to be the low ground to anyone standing on it looking up at the earth) is not a sensible base when compared with deliberately designed and positioned artificial satellites. As a component of a system to accomplish any true-life military assignment envisaged so far, the moon ranges from hopelessly uneconomic to irrelevant. Its orbit is unsuitable. The use of its back side as a place to store missiles and nuclear weapons (which might be expected to survive any attack and deliver massive retaliation) is not necessary and not even advantageous. If we insist that weapons storage locations be extremely expensive to reach and implement and be inhospitable to humans, thinking naively that this equates to invulnerability, such places already abound on earth near the north and south poles, the vast ocean bottoms, and in mile-deep holes sunk into the Rockies.

In military activities, reliable communication between points on land up to several thousand miles apart is a fundamental requirement. The same is true of message transfers among ships, aircraft, and land stations. Satellites can be superior communication relays and sometimes offer the only route for military communication signals. Space offers indispensable military observation points, outflanking the curvature of the earth and exposing all land and ocean surfaces and the atmosphere and space above them to observation. Instrumented packages in space can photograph the earth; probe it to learn what is there; detect communication signals, nuclear blasts, and all manner of radiation and fallout; and provide tracks for missiles and spacecraft. Measurements from space can disclose tests that might otherwise go unnoticed and that might enable one nation to gain a great military advantage over another. Looking from space at what is taking place on and over the earth can distinguish peaceful excursions into space from those that cannot have any but a military purpose. Sensings from space might expose an attack that is occurring or imminent.

Overall, space reconnaissance and communications systems provide substantially greater security during an era when any part of the world must be considered a potential area of conflict and where military force might have to be brought to bear quickly to control an adversely developing situation. Reducing the danger of war by serious arms-reduction pacts is urgent, but such agreements are unlikely to be reached without a system in place that provides continuous information about the war-related activities of the world's nations through indirect (not on-site) monitoring. Adequate inspection and verification of adherence to agreements are hardly imaginable today unless satellite-based sensing is part of the process. Of course, space systems for superior navigation of ships at sea and planes in the air and a similar enhancing of our ability to observe and predict the weather are at least as valuable in military as in civilian operations.

These uses of equipment in space are not limited to strategic warfare applications. In the NATO area, over a smaller but still significant geographical span, the multiple functions of communications, command, control, reconnaissance, intelligence, detection, and warnings are mandatory. Reliance on reception from, and relaying of information between, ground and aircraft stations is by itself insufficient. Satellites in geostationary orbit (rotating about the earth at the same rate as it turns) provide better means to handle many of these tasks.[1]

Land- and sea-based ballistic missiles traveling substantial earth distances pass through and above the nominal atmosphere and into the region of near-in space where satellites roam. When equipped with the right kind of search-and-track apparatus, satellites can follow such missiles and their warhead-carrying nose cones from launch through boost and space travel to reentry into the atmosphere and detonation on their targets. Moreover, it is technologically feasible to fit such space platforms with means to destroy any of these missiles and reentry vehicles that are observed.

Since appropriate equipment in space offers military advantages, removing the enemy's similar equipment from space in event of a war naturally will appear on the military's list of potential actions. In peacetime, space will be populated by apparatus placed there by many nations. If war between the superpowers should come, it must be assumed hostilities would spread to

---

[1] The geostationary orbits are in the plane of the earth's equator. The satellite not only encircles the earth in twenty-four hours and appears stationary to an observer on earth, but more specifically seems fixed over the equator.

space quickly. Some space violence might even precede earth engagements because eliminating warning, reconnaissance, and communication capabilities might be considered essential by the initiator of hostilities. Even nuclear warfare in space must be included in our conjectures. A megaton warhead detonated above the earth's atmosphere would set up an enormous pulse felt by all spacecraft many thousands of miles from the explosion. Unless adequately protected, the electronic systems necessary for control of and communication with those spacecraft would become inoperative.

Whatever a spacecraft's military purpose, it is usually sensible to add coding and antijam features to the craft, despite the great complications this brings to the design, to counter anticipated enemy action in impairing its performance or intercept secret military messages. Thus a military communications satellite typically employs circuit techniques beyond what is needed for a civilian communications satellite (say, for TV or telephony transmission). What with the high cost of the boosting, tracking, and monitoring systems, the payload sent into space by the military ideally should be of extremely high quality, optimized as to function, very reliable, survivable under adverse conditions, and possessed of long life. The cost per pound to orbit a payload is high, so microminiaturization is especially important. Withstanding the launching vibrations and acceleration and enduring the space environment add unusual requirements for ruggedness and thermal and radiation immunity or shielding. Space-based military systems accordingly are extremely advanced in engineering, with everything from basic concepts to design of component parts near the limits of the scientifically and technologically possible.

It is true, as we shall shortly demonstrate, that proposals for civilian space projects are regularly put forth that in boldness, scope, technological reach, cost, and complexity are the equal of any military space project, existing or envisaged. However, military space implementations will often precede the use of similar techniques in commercial areas. Such military space projects serve both to provide for defense-related applications and to push forward the frontiers for eventual civilian space applications.

Space as a Commercial Business

Today's technological capabilities enable the orbiting of equipment for substantial periods, and even the positioning of it at essentially fixed points relative to the earth below. This makes possible the enhancement of certain operations of our surface civilization. These operations include:

Telephony between continents and over great distances on the same continent. Satellites as a link are often more economical and higher in capacity than cables or other alternatives.

TV relaying, point to point, over all the earth by way of satellites.

Airline navigation and traffic control. Satellites act as signaling, artificial stars moving in precise and predictable orbits, in communication with ground computers and airborne transmitter-receivers, making possible continuous, accurate, and economical locating of all aircraft in the skies.

Navigation and communication for ships at sea.

Weather information and prediction. Satellites monitor the dynamic characteristics of large areas of land, atmosphere, ocean, and space and report the data instantly to ground data-processing stations.

Discovery of resources. Satellites scan the earth for data on mineral, forestry, fishing, water, and agricultural resources and pollution, and thus improve opportunities for discovery, warning, and utilization.

Computer-to-computer information transmission between widely spaced points. Satellites help industry maintain logistics, scheduling, and control data and provide access to stored information for professional users.

Direct-to-rooftop TV broadcasts. Satellites, together with local cable systems, bring wideband, multiple-channel programs to mass audiences far and wide.

It is reasonable to assume that in another decade, several hundred million individuals in the non-Communist world will average well over an hour a day in some activity involving a satellite. They may be talking on the phone, watching TV, traveling in an airplane, acquiring needed data at the office, or being instructed at school. All of these services are based upon the workings of an information system partly dependent on equipment orbiting in space. If we put a modest average value of only $1 per hour per person on the space dimension's contribution to making this service possible, the revenues would exceed $100 billion per year.[2]

This range of annual revenue suggests an investment in the same $100 billion range, with yearly returns on the investment around $10 billion (if the

[2] Economic data gatherers would divide this annual revenue of the space systems into such categories as TV, telephony, etc., so the total would not be found listed under "Space" in any GNP figures.

installations are economically sound), a figure greater than the average annual expenditures made in space by the government in the past. It would indicate that we are nearing a period of net positive financial benefit from the nation's investment in space through civilian commercial activities alone, even when national security contributions are not included.

But the real impact of these satellite applications may go far beyond mere financial investment returns. To get some feel of it, consider the role of the conventional telephone in the last century. Without telephony, communication—in the sense in which Americans (especially business employees and teenagers) utilize it—would not have been merely more expensive, it would have been impossible. The ultimate effect of satellites in opening up new communications dimensions will almost certainly be just as revolutionary.

For the commercial applications so far described, the potential market is far from filled, and an increasing number of satellites will be launched annually into space over the coming years. More than a billion dollars' worth of commercially owned satellites are in orbit today. It would appear that before the end of the 1980s, a dozen or more U.S. communications satellites will be authorized yearly by the FCC. There is already strong competition for the available positional slots and radio frequency bands, and future assignments will involve considerable intergovernmental negotiating. Satellites that operate in the same frequency band must be separated because of potential radio interference. (To further complicate the matter of assigning orbital locations, the many underdeveloped nations are anxious not to be dependent on the technologically developed nations for their information-handling requirements and want to share in orbital assignments, even if they may not have the need or the means to exploit the allocations for many years. Some of the equatorial nations even regard the parking orbits "above" them as inherently theirs, an extension of their national territories.) The sum of all this is that as existing applications are expanded, international problems will multiply.[3]

Further-Out Commercial Applications

Considering the relative youth of space technology, new commercial applications beyond those we have just listed will be conceived regularly. A number

[3] In 1985 and 1987, the International Communication Union will hold a world conference and attempt to allocate radio frequencies for satellites in geostationary orbit for the following twenty years. We should not regard their efforts as a failure if they find only a lesser goal attainable.

of ambitious proposals already have acquired enthusiastic promoters. Even if they never reach fruition because of inadequate economic backing, or for any other reason, they are worth mentioning because they show the scope of interest and imagination at work to broaden the dimensions of space.

For example, it has been proposed that huge structures be created in space to capture massive amounts of solar radiation, the idea being to focus this radiation and convert it to microwave radio power so it can be beamed to earth-based receivers. Here the power would be converted again, this time to 60-cycle electricity for use in the electric power distribution network of the nation. Various design approaches have been described to implement this idea. Even those coming closest to economic practicality are still so bold as to astound those with practical experience in engineering and in financial investments.

The National Research Council, in its study of a satellite-based solar power system, said that it would be by far the most costly and complex undertaking, civilian or military, ever attempted, with total costs amounting to $3 trillion (1980) if the system is to provide for a large fraction of the nation's demand for electricity. The space structure would encompass about 25 square miles. Twenty years would be required before a useful demonstration could be made, and another twenty before the system could be brought to full capacity, with expenditures building to the hundred-billion-dollar annual level. An impressive population of astronaut-assemblers would work on the mile-long arrays for decades, and high-payload space boosters would be needed to bring the technicians and the parts into orbit in countless round trips. The unprecedented amount of energy and other resources needed for the construction, the risks, the potential environmental impact, and the numerous severe technical problems to be solved have caused most who have examined the idea to give it a low rating when compared with other alternatives for meeting future energy needs. The government has so far only commissioned preliminary studies. The project may never go further.

Another proposed commercial space activity, also speculative but not so enormously costly and far less risky, is to engage in manufacturing operations in space. Certain classes of materials—unusual metal crystals useful as high-performance semiconductors for computer-communications applications, superior pharmaceutical drug substances, potent chemical catalysts, exceptionally high purity glass, greatly more precise ball bearings—cannot in theory be formed while the forces of gravity acts on the process. In principle, their fabrication could be carried out in the weightless environment of space. This idea obviously would apply only to materials of unusually high market value. Manufacturing in space probably would require a large manned facility there. For certain, a manned space laboratory would have to be developed and committed to this field for years before the potential of manufacturing in space could be solidly evaluated. At this time only some low-cost preliminary experiments are being planned.

Another speculative future space application is a mammoth communications satellite. Here the idea is that because of the high transmission power the satellite could produce and the capacity of its large antennas to pick up weak ground signals and focus return signals on desired spots, very small transmitter-receivers and antennas on the earth's surface would suffice. Such a system could literally put every person in touch with every other person if each had only a proper wristwatch radio, an antenna the size of a soup dish, and a direct line-of-sight path to the satellite. This system is technically possible to build, but no presently envisaged need would justify the cost, so development is relegated to the future, if ever.

Supersized satellites have other possible applications as well. For example, they might carry equipment that could detect tiny amounts of nuclear radiation. This would help pinpoint the location and police the movement of radioactive materials. The satellites could also pick up warning and distress signals and retransmit them to speed help.

Some predict that the relatively small, specialized satellites of today will give way in the future to huge space structures housing many interconnected pieces of equipment, with humans aboard to operate and maintain them, the ensemble serving many purposes and customers in many nations. Also conceivable is a system of highly interlinked small unmanned satellites to provide a world service in the handling of information over a wide range of uses. How this will all work itself out is a matter not only of engineering and economics but of rivalry between the opposing approaches of national domination and international cooperation.

The economic benefits of the commercial applications of space we have listed so far result from putting equipment into near-in space. What about space farther out? What about making money off the moon? So far, by our lunar landings we have eliminated the moon as a cheap source for commercial-grade green cheese. We have learned little to suggest any kind of economic bonanza from the moon's new accessibility. It was claimed by the moon fans early in the space program that it should be good as a bounce point for radio messages from one terrestial region to another, or as an observation point, and would be virtually indispensable as a station on the way to deep outer space. For all these purposes, artificial moons placed where we want them are far superior. We could point to plenty of barren, rocky pieces of the earth much more accessible to us than the moon, and better suited for the support of human life, that remain untouched precisely because we have not been able to see how to gain by populating or removing anything from them. It is quite conceivable that as we learn more, it will merely become clearer that the moon is good for regulating the tides and thus perhaps the love life of the Pacific coast grunion, but not much else.

The more distant planets? Nothing we know about them now would suggest any practical venture for commercial exploitation. Other stars? Perhaps other galaxies? The techniques for getting to them are not even remotely apparent. Those means are more likely to be found, if at all, by indirection—discoveries while pursuing unrelated research—than by a brute-force effort to keep going up and away farther and faster, hoping that the means to reach remote celestial bodies will thereby be revealed. Efforts to breed ever speedier race horses did not lead to the invention of the automobile.

The earth often appears to the human beings on it as an exceedingly crowded place. The world's population probably will continue to grow, and the planet's surface is of finite size. We have built skyscrapers in an attempt to push into a third dimension, but then the two-dimensional traffic problem in getting from home to work becomes worse. Now, however, with space conquest begun, we can dream about the accessibility of an infinite, three-dimensional volume.

The resources here on earth constitute a closed package with a limited amount inside. Even if we learn how to use them to greater advantage, if the population continues to increase at a rate greater than our ability to acquire more resources, our lives may be restricted by the need to divide what we have into smaller and smaller allotments per individual. (At present population growth rates, the resources of the earth must accommodate an additional 60 million people every year.) Space offers the potential of supplies from other planets, moons, and meteors, as well as untapped sources of energy in interplanetary regions. Space presents to us a new frontier beyond which is much more than a new continent or a new planet. We cannot even guess at the scope of it.

As outer space becomes habitable, it will supplant the wide-open spaces we once sang of here on earth. In not too many decades, some earth people may toy with the idea of launching a colony on a planetoid of their own, complete with a farm, a solar-powered oxygen and water producer, and politicians, statisticians, electricians, obstetricians, and musicians. Those who don't like it here could consider organizing to produce a synthetic planet of their own. But this is not the time to invest in any such far-out space possibilities if an early return is the goal.

Humans in Space

Since the United States launched its space program in an emotional reaction to the Soviet Union's Sputnik blitz, it is not surprising that the goal to put humans in space dominated the first round of our activities in this new competition. We simply would not have restored our self-esteem so soon by projects involving unmanned craft alone. The space age was best symbolized by humans heading the conquest, there in person, participating directly.

Sputnik happened to occur just when a related battle over the role of humans was being fought. The arena was the atmosphere, and the contest involved the need to have a pilot in an aircraft. At World War II's end, the most glamorous military man was in the cockpit, a fighter or bomber pilot. But soon guided missiles entered and usurped many important duties, such as shooting down enemy aircraft, delivering tactical and strategic bombs, attacking enemy land and sea forces, and protecting manned bombers against enemy planes. For many key missions, unmanned vehicles are faster, more effective and economical, and safer for the humans involved; they can also operate under a broader range of environmental conditions. Even the flying personnel of the vaunted SAC (Strategic Air Command) did not feel secure as they saw their prestige and prime military function threatened by the arrival of the ICBM.

The first phase of America's entry into space included astronauts very conspicuously, but it failed to establish a substantive continuing role for human beings to play in space. If we want to bring back to earth readings of physical phenomena in space—radiation emanating from the sun or elsewhere in the universe, magnetic or electric fields, density of ions and meteorites—instruments do the basic job, and automatic information-handling devices and communications equipment will bring the information down to earth with greater accuracy and more quickly than humans can. If we want to know how the earth appears from the moon or what the moon looks like to an observer there, electronic devices will pick up anything human eyes in space can, and in greater and more focused detail. The same is true if our purpose is to study the earth for military intelligence and reconnaissance. Also, keeping any spacecraft on the proper trajectory and in a correct, stable orientation is a function best suited to instruments. Finally, if we wish to repair a malfunctioning unmanned spacecraft, then it is true we will have to dispatch an experienced astronaut-technician, accompanied by many tools and instruments and a supply of spare parts, into space to accomplish this. However, simply replacing the entire malfunctioning or worn-out craft through a new launching appears far more economical, especially since, what with steady technological advance, we probably will want to replace it with superior equipment anyway.

The presence of a human being in a spacecraft drastically complicates the project. First of all, the craft has to be designed to return. Providing for the safety and comfort of an astronaut from takeoff to a return landing adds highly penalizing weight, cost, and time to the exercise. It narrows the range of permissible risk taking. For a project involving ordinary unmanned space-craft, for example, in which a group of ten vehicles might be contemplated, if one of the ten should fail, it would represent the equivalent of a 10 percent cost increase in the program. However, giving an astronaut only nine chances in ten of returning is totally unacceptable. Yet for some projects a 90 percent probability of success may be the highest reliability attainable at any but prohibitive costs.

Whenever a system is being designed, whether in space or on the ground, a competent systems designer will try to choose the right combination of humans and machines to accomplish the given task sensibly. To rely totally on human hands and backs, or brains and senses, and provide no associated mechanical or electronic tools is usually an extreme, not the optimum. Conversely, it is rarely best to completely automate everything, because people are produced by relatively cheap labor, entail reasonable annual maintenance cost, and have some talents hard to duplicate with a machine. Thus a person can reach into a basket of odd-shaped parts and pull out a square, round, or star-shaped object, quickly identify it as such, and place it in a similarly shaped hole. For a robot to do the same simple task of selecting, identifying, concluding, and fitting would be prohibitively costly and complex. On the other hand, mechanical devices can exert tremendously greater force than humans and withstand a much more severe environment. A human being can multiply a one-digit number by another at the rate of approximately one multiplication per second. A computer can multiply two numbers, each with many digits, in a millionth of a second. People and machines each have their places. In space, as on the ground, it is wise to break down any task to be performed into its components, with an open mind initially as to the role of the human being.

If this assessment is done for civilian and military applications in space, few essential functions requiring a person emerge. The Apollo flights, which featured astronauts, came to a dead end when enough successful landings had taken place and the program had accomplished its psychological mission. Public interest waned and NASA budgets drifted downwards. Civilian space applications, sans humans, received all the attention of the private sector, and the need for unmanned satellites took first place in the military's priorities.

Pure research in outer space continued, but attention turned to the more interesting exploration of other planets with unmanned, instrumented spacecraft. Some irrepressible promoters propagandized for a manned Martian landing program, but this trillion-dollar possibility received more smiles than support, and a modest manned laboratory program that accomplished many firsts in understanding man's abilities in a space environment was canceled.

An instrument package was landed on Mars which, without the presence of human hands and senses, scooped up Martian surface matter. Then highly automated, microminiaturized laboratory and computer-communications equipment within the package processed the material, examined it carefully for signs of existing or past life, and communicated the results back to earth. The rapid advance of information technology is steadily making possible more sophisticated automation in information handling, whether it be sensing, processing, on-board control and navigation, or communication back to earth. All of these functions can be done at increasingly less cost, with less weight, and with increasingly high reliability. This militates against the need for humans.

Until the program to develop the so-called space shuttle began, the United States had accepted a seven-year hiatus in placing humans in space. The shuttle arose out of two major influences.[4]

One was economic potential. Given enough payloads to be orbited regularly, it seemed it might be cheaper at some point to do the lifting by a system in which at least part of the boosting equipment returned to earth like an airplane, to be used again and again. Furthermore, the shuttle was big in design, capable of taking up a large number of payloads at the same time; thus some common launch costs could be shared. A reusable launcher like the shuttle, capable of carrying a manned crew, the main equipment designed to survive many launches, was known to be an expensive system, but it was believed that it would work out in time to be cheaper than nonrecoverable launchers after enough repetitive voyages. Expendable boosters were regarded as obsolescent.

The other influence was the growing pressure to reinject human beings

---

[4] Some observers, highly critical of NASA at the time, felt that there was a third influence affecting the shuttle program. It is that NASA leadership yearned for the good old days of manned lunar landings and desperately felt a new, grandiose program to be essential for the agency to survive. These critics, some still vocal, believe the shuttle should have been a nonurgent program and did not deserve its priority and large-scale funding.

into the space environment. The unsatisfied feeling remained that ultimately space must be added to the habitable regions of the universe. Exerting some weight were potential future projects, some described earlier in this chapter, which would require the presence of human beings in space for their assembly and operation.

Thus the shuttle project commenced. But the program planning and selling were dominated by an unfortunate malady that has plagued many large military programs. The project was oversold, and accomplishing it was underestimated. First, the guess about the market was wrong. The era of frequent launchings of large weights was anticipated as arriving earlier than now seems the actuality. Second, the engineering difficulties of the shuttle and the time required to complete the design and testing were preassessed too optimistically. Consequently, inadequate funding was allotted. As has happened again and again with military programs in which technological difficulties, time to completion, and overall funding were portrayed too optimistically at the inception of the program, schedules slipped and more money had to be begged. The total payload the shuttle could carry had to be revised downward, and the cost per pound to orbit payloads adjusted upward, even as the turnaround time and necessary repairs after each launch were seen to be beyond the estimates.

The program is not yet complete, and it is natural to expect more engineering problems, slippage, and additional funding increments before the shuttle will really be ready for reliable, repetitive use. Uncertainty about the availability of the shuttle to boost important military and commercial satellites into orbit on planned schedules has created problems and embarrassments. For example, some commercial American satellite projects originally scheduled to use the shuttle have been rescheduled to use a new European-made (largely French), nonrecoverable booster—the so-called Ariane—which offers launchings at a lower price and appears able to meet the schedule requirements.

One budgeting advantage originally claimed for the shuttle was that since it would reduce launch costs, it would make possible more launches and a fuller overall NASA program. Actually, the increased cost to maintain the shuttle program has caused the cancellation of other programs, such as deep-space research missions. Here the United States had been the leader, and we were forced to abandon joint programs with other nations on which they had already expended considerable funds. The Russians, French, and Japanese will rendezvous with Halley's comet as it nears earth—without us.

The shuttle remains a controversial project. Some argue that, even putting aside the initial misestimates in development time, funding, cost to launch, turnaround time, etc., a modest research program would have been more appropriate anyway. If found to be a sound idea through such a program, the shuttle could have developed into an operational project later. Meanwhile, the continued use of proven nonrecoverable boosters would have saved money and added reliability to the launching dates of urgently required spacecraft now awaiting the shuttle. We would not have seen American boosting business shift to foreign competitors, sad events in view of the enormous American space expenditures and pioneering efforts.[5] Again, if the shuttle program had not been influenced by a desire to inject humans into the mission and by a premature interest in accommodating the highly speculative, supersized space structures of the distant future, the shuttle's physical design could have been simpler. Even granting merit in the idea of recovering and reusing part of the booster system, the entire design might have required less time and funds if the shuttle had been designed for unmanned flight.

On the other hand, surely an American space program that assumes no role for human beings in space cannot be counted as acceptable. We cannot presume to know today all of the applications that may become important. Thus we should develop enough know-how to be able to add human actors to space activities without undue delay if their presence becomes the key to an unforeseen mission.

And then there are always the Russians. They seem intent on maintaining a manned space station in orbit. Their program's progress will force continual questioning of our space program's interest in human passengers. The Soviet Union's reasons for placing humans in space, whatever they are, may be less significant than the single fact that they are carrying on such activities—because this always will suggest to some observers that they are ahead of us. Merely in reaction to unfavorable comparisons, we might feel compelled to build a permanent U.S. space station, as NASA highly recommends we should. The real rationale, if we go ahead with a permanently manned space platform, should be to do research on a human's ability to

---

[5] Competition from other nations in space is inevitable and has its benefits. Moreover, the field will develop faster if several nations make separate and different contributions as well as cooperating regularly on individual projects to share costs. But the Ariane booster fills a need for a straightforward expendable booster on which the U.S. program simply defaulted after developing and producing a large number of essentially similar equipment.

operate in space and to perform experiments (for example, on the key aspects of manufacturing in space) which only a human can do.[6] It should be sold as a research program. It should not be pictured as a high-priority must and tied to national security needs said to be essential but actually unidentified and truly, as yet, unknown.

Something also needs to be said for the pioneering drive of humankind, and the way it can be ignited by participation in bold, pioneering space events. The Apollo moon project serves as evidence of that phenomenon. In the economically troubled times of the early 1980s, space efforts will get little priority if they are seen to offer no more than intangible, spiritual gains. Such realized satisfactions are worth something, however, and when we are forced to conclude we cannot afford to generate them, this should be counted as disappointing. The moon may be boring to visit again, and there is absolutely no indication of either critical usefulness or epoch-making romance in a habitable platform stationed over the earth. Of course, in some future, more challenging time, lonely Mars might beckon with irresistible allure. Men and women from earth landing on that distant neighbor—that would be an event we would all stop what we were doing to watch on TV. But it hardly calls for starting a crash program. Not now.

Space as a Research Frontier

We have managed to learn a great deal about the universe from observations made on earth, a highly specialized and insignificant little piece of the whole. By moving into outer space, we can escape from the shielding and interference of our atmosphere, the magnetic field, earth-emitted radiation, and gravitation. We can open up new closets in which are stored some of nature's secrets totally unknown to us in our own little isolated pigeonhole. Some even believe that one important research quest in outer space should be to

---

[6] Another substantial reason why a program that leaves human beings out cannot be judged perfectly balanced is the unique opportunity space provides for research on the nature of biological matter and, most particularly, on the human body. Where there is no gravitational pull and where other environmental factors are different from conditions on earth, the body's main organs and systems will surely function somewhat differently. Observing these differences may lead to an understanding of the living and aging process in humans that might otherwise escape our notice and remain beyond our imagination.

determine whether intelligent life exists elsewhere. Of course, it is not certain that we will discover the answer sooner by sending people and equipment into space looking for it. We might do as well by staying here and merely watching diligently for signals sent a long, long time ago by other intelligences to let the rest of the universe know of their existence. But all that we ever can hope to learn about the universe should become increasingly accessible as we move away from the confines of our planet.

Every aspect of nature, including the atoms and their tiny constituents available to us here on earth, involves a boundary between what is known and what is still mysterious or completely unperceived. In this regard space is not unique. It is merely one frontier deserving investigation, and the attention we pay to it should compete with study of numerous other frontiers. Still, we perform some research tasks ahead of others because for those chosen ones we happen to have the means, we are ready, and the investigations can be done at reasonable cost, whereas other probings, just as fascinating and potentially as ripe for important discoveries, are not so readily performable with presently available tools.

With unmanned spacecraft we can now track down the main characteristics of the entire solar system. This includes the makeup of the planets and their moons and rings and their highly peculiar surrounding environments. We already have extended a communications net into deep space. Doing so has brought us pictures of Saturn and Jupiter from distances of over 1 billion miles. We have the ability to map the surface of Venus, which is totally clouded over and hence not observable here on earth. We can establish spacecraft on trajectories to rendezvous with comets. We can continue considerable space research without developing entirely new spacecraft; the designs we have already built have been proven in journeys to the edges of the solar system and can be modified as required for new missions. Observations of emanations of the universe beyond our sun and its planets also can be made economically now from new vantage points which our solar system affords.

The United States has exerted world leadership in exploring the solar system. The visitations by our spacecraft to the outer planets have been among our century's most inspiring scientific and engineering accomplishments. If we do not exploit the position America has attained in outer space, we will be throwing away a large part of a major investment. Moreover, probing the universe's natural laws from space regions never before accessible to scientists might well lead to discoveries we cannot even anticipate because they are pure unknowns. If there are such possibilities, they have implied military applications. We cannot with great comfort allow potential enemies to gain such new knowledge for their exclusive use. Space remains a new field in which to prospect; it is far too early to leave the finds to others by default.

To do right by the nation in our space research programs seems to require a far better organization of the scientists at large, in all fields, so that allocation of funds to the various fields—biology, geophysics, oceanography, and all the others, as well as space—will be the result of deliberative statesmanship by research leaders. Scientists are reluctant to organize themselves for their proper and necessary role of educating the public as to the need to pursue—and the benefits of pursuing—scientific fundamentals, and they have even less enthusiasm for setting themselves up to choose which fields to finance generously and which to explore lightly or defer. The prestigious academies wait to be asked for their analyses or opinions.

The political arm of government has a sense of how much of any given year's budget should be devoted to all research—but it is only a political value judgment, formulated with today's procedures and applied with a minimum of scientific judgment. Specialized groups of scientists, and often individual researchers, seem interested only in receiving the funding for their own work. In this the scientists pushing their projects act much like corporations marketing their contract proposals. But this means space research, or any other field of research, becomes too dependent on promotional skills or on the short-term happenstance of budgeting. This condition will continue until the nation's scientists broaden their endeavors to include organized political lobbying, active statesmanship, and public education as to the role of scientific research in the nation's future.

Free Enterprise in Space

It is now over a quarter of a century since Sputnik; still space is not a mature business field. If we wish to ponder the best missions for the government and the private sector in the application of space technology, it is well to keep this continued newness in mind. For instance, a large commercial business based on communications satellites is building, but even this application is dominated by the rapid advance of technology and is not yet well established. Again, because of the large military space technology programs and the changes government is forcing in the means for boosting spacecraft into orbit, government-sponsored research and development exceeds the efforts financed by the private sector and keeps shifting the base on which free enterprise risks can be planned. Finally, considerable cooperation must be arranged among national governments of the world for successful continued growth in space technology applications. This international dimension is a dynamic one, requires strong U.S. government participation, and thus greatly affects the private sector's role and potential.

To explore the value of free enterprise in space, let us consider communications satellites more particularly, since these come closest to being ready for delegation to U.S. private industry. Despite competition and suspicion among nations,[7] an international telecommunications satellite agency, Intelsat, has been running quite smoothly for years and now has over 100 nations as members. Intelsat provides over 20,000 full telephone circuits, which is about two-thirds of the world's total transoceanic telephone circuitry, telex, and data links. The world's civilian market for communications satellites is gearing up to meet the non-Communist world's plans to orbit some ten to twenty satellites per year during the 1980s. This activity, together with associated ground installations, adds up to several billion dollars of annual expenditures for the R&D, satellite design, production of hardware, and costs of launchings (as distinct from revenues from the services which those installations make possible and the investments in the conventional ground-based communications networks, such as telephone or cable lines, into which the satellites tie).

This is a large enough volume of business that a number of space engineering organizations can be expected to serve it, including Japanese and west European as well as American participants. As it reaches the level of several billion dollars a year, this program should support over $100 million of annual research and development, which will advance the art steadily. Thus what is done in the following years will transcend earlier efforts in a pattern of mutually reinforcing technological advance and revenue growth continuing for decades.

For some aspects of U.S. commercial satellites, the government is now essentially out of the picture. One can argue that had superior leadership

---

[7] Communications satellites offer economic advantages for all nations and often perform across national boundaries, a function requiring international agreements. National governments attach importance to control of communications directed at their countries from beyond their borders. Just as nations do not want others snooping from above, seeking to learn about their activities and resources, they also do not want direct broadcasts coming from foreign countries to the antennas on the rooftops of their citizens' homes. They feel that messages that enter in that way can influence the thinking of their people, pull away advertising revenues, and perhaps sell their citizens foreign-made products or unsell them on their own governments.

existed in private industry twenty years ago, at the time communications satellites became technically feasible, the field could have been a free-enterprise operation right from the beginning. The larger communications companies in the United States (AT&T, IT&T, Western Union, RCA, etc.) could have justified adding the space satellite dimension to their communications facilities then, as could have other electronic or computer companies who even in that early period should have seen that communications satellites were directly related to their future growth. Leaning heavily on the technology already developed for boosters, satellites, and launching and tracking facilities by government projects such as the ICBM, the Apollo, and military and research spacecraft programs, these companies would have had to take only a reasonable investment risk—say, in the range of $100 million to $200 million each, their net worths being in the billions—to have prepared to add satellite-based telephone and television broadcasting systems to their services.

The government preempted this private approach and took the initiative in establishing the first communications satellite company, Communications Satellite Corporation. (Starting as a virtually government-controlled company, Comsat later became a publicly owned corporation with only a part of its board of directors still appointed by the president.) Left to themselves, the existing communications corporations of that period would have gotten started, but much later, since their managements lacked imagination, boldness, and appreciation of the potential of satellites. Thus time was certainly saved as a result of the government's dominant role in the beginnings of commercial communications satellite systems.[8]

While much of the initiative for civilian communications satellite projects in the United States has now passed into the hands of private corporations, the government remains powerful in this area. Every communications satellite system proposed by the private sector, such as by Comsat or Western Union, must be allocated operating frequencies and bandwidth in the limited radio spectrum, positional slots in geostationary orbit, and patterns of coverage of the earth below. All these are controlled by the U.S. government for American corporations. Moreover, what the U.S. government can allow its corporations to do has to be consistent with international agreements that divide up these entitlements, since arranging such agreements remains a government function. Finally, the satellites must be placed in

---

[8] The government's action was itself unusual, the result of NASA's good fortune in being led at the time by a remarkably competent and farsighted director, James F. Webb.

orbit, and the U.S. government still controls the physical orbiting of all space objects launched from this country.

Of course, it would not be illegal for a private American entity to procure appropriate launching equipment by contracting separately with companies that manufacture the large rocket engines and the booster structures. Even though launching technology relates to military boosters and such other associated classified apparatus as ICBMs, the U.S. government would allow such independent booster activities to develop. Admittedly, the art in boosters is constantly advancing as a result of the government's programs, and the resulting changes influence what private groups will choose to do. The shuttle, for example, is altering the U.S. booster business greatly, and altering as well the potential for independent entrepreneurs who might choose to enter the field as private providers of boosters. A private U.S. booster company would have difficulty procuring any U.S. government space business, even though its quoted price to loft a satellite might be lower than the cost of using the shuttle for many of the payloads now envisaged. To make full use of the shuttle, the U.S. government may be expected to insist that it be employed in the future to launch all government spacecraft and satellites, military and nonmilitary. However, had a U.S. entrepreneur decided to offer a nonrecoverable booster, as the Europeans have done, we know now that other U.S. companies interested in orbiting commercial satellites would have been prospective customers, since some such business has already gone to the European competition.

If booster equipment and the satellites that they orbit were totally available through U.S. free-enterprise activities, the need would remain for carrying out the actual launching and tracking. Launching facilities, at the Kennedy Space Center in Florida and Vandenberg Air Force Base in California, are government-owned and -controlled and were extremely costly to build. The tracking ranges involve a complex network in space and installations both on U.S. soil and in other countries arranged through government-to-government negotiations. These installations, created first for the ICBM program and then extended for space programs, were developed over a period of a quarter of a century. It would cost billions of dollars to duplicate them, even if arrangements could readily be made again for suitable land in just the right places in the United States and around the world. This would be prohibitively uneconomic for a private group. Moreover, the present U.S. government facilities are adequate and can readily be extended to handle all envisaged traffic. Communications satellite projects, even those totally funded by the private sector for civilian applications, if launched from the U.S. facilities, have to fit their launching dates into a government program whose prime interest is military satellites and other government-funded spacecraft. New factors arising suddenly on the security front could raise the priority of military spacecraft and push civilian space projects back with no notice.

We conclude that even in the commercial satellite field, where the private sector is now closest to playing the lead, the U.S. government's mandatory involvement limits free enterprise. Of course, the government makes use of the free-enterprise sector through contracts for R&D, procurement of military communications spacecraft, the development and operation of boosters, and the maintenance and extension of launching and tracking facilities. Despite the permanent presence of government, the principal drive in expanding commercial satellite applications in all its dimensions in the United States should henceforth come from the private sector. The computer-communications industry, it is to be hoped, will build up, not diminish, its entrepreneurial zeal. It should move ahead aggressively on its own in space, and it should push the government to provide support as required where only the government can. The industry will let the nation down badly if it reverts to the hesitant conservatism that characterized it two decades ago, when government leadership was needed to create the first corporation to exploit the civilian use of satellites in communications.

Let us shift now to the role of free enterprise in the military space field. Military satellites for command, control, communications, intelligence, and reconnaissance are among the highest priority and most demanding of military technological projects. However, this is not a field for true free-enterprise investment because the market has a single customer. The pattern is now well established, and it is quite like all other classified military technology programs. Risk investment, made to seek a head start in acquiring future military business in space, can be expected to remain extremely modest in scale. Technological corporations will happily contract with the government to design and build military spacecraft, but will not invest large amounts of private funds to do so ahead of a firm contract.

The same applies to projects involving space probes to Venus or Jupiter, or rendezvous with comets, or platforms in space for sensing light, radio signals, or x-rays from distant parts of the universe. Space research is not an ideal field for private investment. Learning more about the laws of nature by making observations in outer space is not an activity that an American corporation could justify funding with its own capital, even if extraordinarily long-range thinking characterizes its management. The results eventually should benefit all the people of the world. If the United States wishes to make a contribution here, it is a fitting area for funding by American taxpayers as a whole. In fact, in contracting for pure research in space, as distinct from classified projects, it is particularly sensible for the government to favor the use of universities, rather than of profit-seeking private industry. The graduate studies that lead to the higher technical degrees necessary for performing research and development in both industry and academe are best undertaken in an environment in which the nature of the universe is being constantly explored.

The Unsettled Government Mission

The assignment of roles to government and to the private sector is much less settled for certain other important applications of space technology than it is for communications satellite systems. One such application is airline navigation and traffic control through satellite systems. No such system can be brought into being, it turns out, without the participation (equivalent to a vote or a veto) of a very large number of semiindependent entities. These are the airport operators (cities, counties, and military services), the airlines, the Federal Communications Commission (FCC), the Federal Aviation Administration (FAA), the Civil Aeronautics Board (CAB), the pilots' associations, the Army, Navy and Air Force, the National Aeronautics and Space Administration (NASA), and the companies making the satellites, radar, airborne computer equipment, and ground-stationed apparatus. To make the system work, it is necessary that there be rules requiring every airplane to carry appropriate equipment and thus to cooperate with the system. Since foreign planes must operate in our environment, foreign governments must also be involved in specifying the system.

Clearly, there are numerous potential roles here for units of the private sector possessing the required expertise. They should participate in studies, design proposals, research and development, and finally in the production of hardware and software to make the entire system operable. However, in the overall scheme, this cannot be a free-enterprise project; the system responsibility is appropriately placed firmly with the government. Unfortunately for space-based airline navigation and traffic control, our experience with other applications of advanced technology in which the government must be the coordinating player demonstrates that we do not have in place a suitable government organization. No single entity within the government has the responsibility to see this job through from beginning to end.

No law prevents the government from setting up an appropriate systems management unit within the FAA, for instance, using perhaps an industrial systems contractor to provide it with the necessary systems engineering, integration, and direction skill. (This was done successfully by the Air Force on the ICBM program.) Lacking the organization to get the system designed and implemented, we have let decades pass without yet having developed and put in place the computer-communications equipment and appropriate satellites to constitute a space-based airline navigation and traffic control system, even though it would be eminently sensible from the standpoint of improved economics, traffic capacity, and safety. The first step is for the government to recognize the void and organize to fill it. After the proper organization is created, the problem will still remain of doing the job well and enlisting adequate support from the free-enterprise sector. Step number 2, implementing the system, will be difficult, but step number 1 has not even been started.

The government's role is better established in the weather prediction field. Weather satellites have been operational for about fifteen years and are now basic to weather forecasting, providing daily TV weather maps, and improving operating decisions in transportation, agriculture, fishing, and other fields. Here the National Weather Service had a ready-made infrastructure in the United States when space activities began, and no problem of national sovereignty has arisen. The service was capable of setting up for dissemination of data from weather satellites and was allowed to do so when the space dimension was opened. (Amazingly, some things happen as they should with our government.) The World Meteorological Organization, in existence for many decades, has been able to coordinate the interchange of data worldwide. The pervasive use of weather data makes the obtaining and distribution of it a natural, probably permanent function of government.[9]

However, another example, the Landsat project, illustrates the puzzling government-private policy questions to which the applications of space technology can lead. Landsats are satellite-based systems that examine the earth and show promise for mineral exploration, water and forestry resource evaluation, pollution monitoring, the assessment of crop and soil conditions to aid agriculture, mapping, snowcover analysis, beach erosion studies, and more. Suppose a private American corporation were to orbit a Landsat as a free-enterprise project. This company, let us imagine, processes the data that

---

[9] This discussion does not cover the separate (and classified) weather satellite programs of the DOD, which cover the military's considerable interest in weather data and prediction.

results from scanning the earth's surface and looks for valuable mineral finds. By marketing the information, it expects to reap financial rewards that will constitute a favorable return on its investment of several hunderd million dollars.

The company would need allocated frequency bands for communication, government approval of the orbital trajectories, and, for an appropriate fee, the use of government launching and tracking facilities. These steps are the easy ones to arrange. Privileges the company would seek as it tried to exploit the acquired data would raise with the government the question of whether it was proper to allow a private corporation to engage in Landsat activities. It would not be easy for the U.S. government to make this policy decision even if it alone were involved, but the problem is compounded by required intergovernmental agreements.

How would the risk-taking company obtain revenues? Should it sell the raw data to all comers? Suppose it processes the data first and they indicate the possibility of a previously undiscovered valuable resource in a foreign country. Should the company go to the government of that nation and offer to tell what it has learned for a price? Should it demand a percentage of the value of the minerals that might be extracted?[10]

How does the country thus propositioned even estimate the worth of the information? It must first be provided with the information, perform analysis on it, and follow this with detailed on-site investigation of the scanned terrain—after all of which, why should it favor paying anything? Perhaps, as in a class-B movie, the company should offer to sell the information to a private clique in that country, which would then secretly buy up the pertinent land or the rights to the resources on it (assuming the country allows such private ownership).

Even if the data should apply to a minerals find in the United States, a policy problem would assert itself immediately. Of course, in anticipation of the utilization of this new technology by the private sector, the U.S. government could create laws to determine the rights, privileges, and appropriate compensations for private groups who, having made investments

---

[10] As a rule, countries jealously guard all information pertaining to their natural resources. There is also an unfortunate relationship betwen Landsat and spy satellites. Distinguishing details of structures on the ground and identifying them as either military or civilian in nature are accomplished by employment of highly classified techniques. However, these techniques may not be very different from those that make possible the discovery of ore-rich regions.

in the new technology and gathered valuable data, presumably should be entitled to a fair return. Perhaps, as now happens with oil exploration on land or offshore, the Landsat company might simply use the information quietly to acquire land and rights that would not otherwise be appreciated as valuable. But there is more to it than this.

Observations of the earth—sensed by a Landsat and then studied by a partnership of computers and human experts—can disclose conditions of the earth's surface valuable for agricultural planning, controlling crop disease, finding and utilizing water resources, indicating mass pollution effects, anticipating and measuring the severity of flood conditions, and other public service data. What do we do about these data whose dissemination would appear to be in the public interest and not likely to be adaptable to proprietary exploitation for profit? In fact, it is certainly in substantial part to achieve such potential gains for society that the government has spent a billion dollars in research on the Landsat concept.

The technology is not yet adequately developed and proven, and it is not clear that the promising possibility will turn into a system that regularly uncovers valuable finds. For a long time corporations dealing in petroleum and minerals may believe it more sensible to put their available financial resources behind more conventional exploratory activities. It may be that only when and if the first important discoveries are made, will it be the right time to consider how best to move the program along commercially. Meanwhile, the government will continue to foot the bill to do research in the scanning techniques and the analyses of data. Landsat today is simply an exploratory research and development program and as such is properly lodged in NASA.

How the United States should handle Landsat as an eventual operating system is unfinished business. It is to be emphasized that this is an application of space technology which might be of tremendous value to the nation and the world, and some of the potential appears suitable for implementation by the free-enterprise sector. In America we can really expect to profit most from injecting the free-enterprise concept into the act for stimulation, generation of incentives, risk taking, and anticipated appropriate reward. Failure to use free enterprise in Landsat activities would be like expecting to explore for oil and minerals only through government-sponsored searches. Yet in the short term the government must dominate the whole Landsat idea from R&D through to control of dissemination of obtained data. The government must be in the chairmanship position, integrating the whole and piecing out appropriate parts to the private sector. No government unit today has the broad charter to follow through on Landsats for the long term, to search out the most useful data, disseminate it for maximum exploitation, determine the private sector's role, and arrange for implementation.[11]

Landsat is not the only space application still to be organized as to the roles for government and for free enterprise. The ocean-observation satellite is an example of another such application. It displays the conditions on and in the oceans, regions of the world largely beyond national jurisdiction. No institutional structure exists today for operating a civilian ocean satellite system, yet much might be gained for the world in acquiring and distributing data about transport, fishing, and ocean resources and phenomena. (The colors of reflected radiation from the ocean from various aspects disclose the nature of biological life there and can be used to direct fishing expeditions to the best areas, with consequent lowering of costs and energy consumption in commercial fishing.)

As another example, NASA is just beginning to grapple with how to turn the shuttle into an operational program once the R&D phase is completed. To what extent should free enterprise be involved in taking over the shuttle as a routine launcher of everyone's payloads for a fee? Here again, U.S. industry should be more aggressive. Once before, it stood by with helpless caution while the government deprioritized expendable boosters, allowing the European Ariane to fill a gap that U.S. industry, with its enormously greater experience, could much more easily have satisfied. Today the industry is still quite inactive in inventing and setting its own preferred role in eventual shuttle operations.

The Triangle and Space

After twenty-five years it is still true of the entire commercial use of space in the United States that the government and the private sector have not yet worked out their best permanent roles. Less forgivable is something else. With space so clearly an arena of powerful economic and security interest for the nation, we have been approaching plans and policies about space for well over a decade on an intermittent, hop-and-jump short-range political basis. NASA has many hopes and plans, of course, but the nation does not have a plan for the next two decades. A real plan would describe both goals and

[11] It may well be that after amendment of the U.S. antitrust laws (as urged in the next chapter), a number of companies might form joint ventures to create commercial Landsats and to carry on beyond where the government, with its lack of expertise as to real market potential, should be allowed to go.

anticipated budgets. It would have recognition, acceptance, and stature with all the power centers influencing advances and applications in space, namely, the government's executive branch, Congress, industry, and the scientific and technological fraternity. A real plan would be one to which all these forces were committed long-term, in the same way that at the start of the 1960s we were committed to landing a man on the moon before the end of the decade.

Why is this? Must we be without a plan? The triangle of society-technology-liberty certainly imposes no forces on space research that end up distorting objectives and inhibiting progress. True, the ugly possibilities of space warfare and space weapons races suggest again that the society of nations, if it wishes to fully realize the benefits of space, must consider agreeing to restrictions on national liberties to act unilaterally in space. Also, economic constraints of several kinds exist to limit the scope of space research and the speed of implementation of useful applications of space technology.

But none of these factors should prevent the United States from having sound long-range space goals as a guide to the government's budgeting process or to the setting of its own roles. Neither do they preclude our laying out sensible missions for free enterprise in space, turning over appropriate areas to the private sector, and encouraging private risk-taking investment through consistent government policies that create the environment for such investment. It is rather that the typical American operating pattern stands in the way of progress in space. Government, private industry, and the scientific and technological fraternity are equally responsible. Each of these participants exhibit shortcomings in the way they go about things, shortcomings that are emphasized in the nation's space program.

Space research simply presents a highly conspicuous and highly magnified image of the good and the bad in the way technological advance proceeds in the United States. Space exploration was born out of an emotional reaction, the first basis for an exaggeratedly up-and-down future. Government, industry, and scientists and engineers have all had a hand in the accomplishments of the space age, but they have also contributed to the weaknesses in organization, decision making, and implementation. Less-than-adequate attention has been given to setting priorities and long-range goals and allocating missions to each sector. The participants in space, government and the private sector, have to depart from their self-seeking, narrow efforts and give more attention to the decisions and organizational problems that affect the overall national interest. Then the useful dimensions of space will be expanded.

Chapter Seven The International Battle for Technological Superiority

It would simplify things if the United States were privileged to pursue the goal of using science and technology beneficially, with no outside interference. Of course, it would not impair our efforts if we could tap other nations' raw materials as needed, incorporate scientific discoveries and technological know-how developed elsewhere when ahead of our own, and be compensated generously for products of ours we chose to sell. Acting on such privileges could only enhance our profiting from science and technology.

But the rest of the world does not exist to serve us. Moreover, there is more to international relations than seeking a maximum utilization of science and technology. Communication and trade between the United States and other countries—the motion of people, goods, money, and knowledge—are in response to understandings and customs that range from carefully thought-out and formally negotiated arrangements to happenstance. The interfacing of nations occurs in only partial order and considerable confusion.

As with every other aspect of the society, international relations are affected greatly by scientific and technological advance. No international issue—trade balances, cold wars, energy, import barriers, influences of multinational companies, cartels—is without its scientific and technological facets, and they are sometimes dominant in creating the problem or opportunity and determining how it should be handled. As we observed with domestic issues, we cannot isolate the scientific and technological dimensions of international problems for separate handling. But we can profit from asking how international issues impinge on the objective of using technology most sensibly in the national interest and what changes in our policies might help attain that goal.

Sound use of science and technology by the United States is necessary for more than domestic tranquility. Our country is not to be seriously described as a competitor to the underdeveloped nations in providing low-cost labor or in producing nontechnological products for consumption by the more industrialized nations of the world. America's position, if we are to make the most suitable contribution to world economic stability, must be that of a strong, high-technology country. Even our trade surplus in raw foodstuffs depends on our technological position, because without our machinery and chemicals we would not produce gainful agricultural exports. If the world's economic soundness depends in a major way on each nation's doing what it can do best, making optimum use of its human and natural resources, trading portions of its output to others, and receiving in return what those others are best able to provide, then the United States must offer high-technology products to the world.

Because a technological base has become influential in the economy of every nation, technological advance is now a critical arena of international contesting. The competition is as ferocious as is to be found in any cold war. The fight here is for favorable positions in technological product development, improved manufacturing processes, and superior distribution systems. The leaders of governments and technological corporations now find the battle for technological supremacy to be a principal subject of policy and strategy deliberations. In fact, it is the government-industry interface as it pertains to international competition that often holds the key to unlocking advantageous employment of technology. In the United States we have difficulty getting the key in the lock and turning it all the way. As we shall see shortly, to enhance our possibilities of leading in technological advance will require innovative twists in relationships between our government and the private sector.

Our Vanishing Captive Domestic Market

Decades ago our broad technological superiority showed itself in the high worldwide regard for American-made consumer and industrial technological products. The large American domestic market itself was almost a captive one for our home-based industrial corporations, and our products were generally the world's most advanced. During that period producers in western Europe and Japan could design products as good as the American ones in some high-technology fields and had a lower wage rate applying to all. However, our quality and productivity were usually higher because of greater investment, a broader infrastructure of skilled human resources, more modern facilities, and the accumulation of more advances in engineering design and manufacturing processes geared to our higher volumes: thus the final performance or pricing on technological equipment of other countries usually could not match ours.

In fact, during the first three decades after World War II, most of the world thought it necessary to create high tariffs on imports from America in order to give their own technological industries a chance to establish themselves in their internal markets. They rejected the alternative of going on working at poorer wages, producing lower-technology products to sell in the American market for American dollars needed to pay for higher-technology imports from us. Instead, they protected and subsidized their own technological industries.

In that era, whether we were employing science and technology less than perfectly was not decisive, and we could be sanguine about our domestic market. If the allocation of missions between the government and the private sector was overly casual, our highly advantageous position against world competition masked that shortcoming. As to military technology, we were well ahead of the Russians then, and American leadership, technological as well as economic and political, provided most of the initiative for the military alliances of non-Communist nations. Finally, during that period we were close to self-sufficiency in energy and other natural resources, largely buying those from outside, which we did regularly, only when we could do so more cheaply.

Those days are now well behind us. The improved quality and favorable pricing of the output of foreign producers have enabled them to move successfully into the U.S. market. Although we have more physical resources than western Europe and Japan combined, their ability to put world resources at the disposal of their technological industry is equal to ours. Suddenly some Americans are heard saying we should adopt the approaches other nations have used in the past, when they were inferior, and that we should now subsidize our technological industry and erect barriers against the import of technological products.

But do we have this choice in practice? If we are being outdone in some technologies by western Europe and Japan, then banning entry of those products would raise our prices. By insulating ourselves from the benefits of their efficiency and advances, we might get further behind. Of course, keeping the large American market walled off from other nations would handicap their progress. Beyond question, both camps would suffer. Higher prices and lower standards of living would result everywhere.

In choosing what to do, we cannot ignore our battle against inflation with the possible undesired accompaniment of recession and unemployment. Japanese steelmaking is now characterized by more modern technology and plants, higher productivity, less energy use, and lower prices; but if we were to ban both foreign steel and foreign automobiles, this would hardly guarantee that American automobile manufacturers would return to earlier levels of employment. With steel more costly, the prices of American-made cars would rise. Our car manufacturers would have all of the American market, but it would be a smaller one. The industry could not become healthier than it is today if few Americans could afford a new car.

Attempts to deny entry into the U.S. domestic market of products made by foreign countries also would collide with national security requirements. We could not place an embargo on technological imports from western Europe and simultaneously maintain an alliance with them to deter Soviet military aggression. The decreased exports would quickly cause horrendous economic difficulties for those nations, which would soon experience social and political upheavals, a lowering of their already modest enthusiasm for defense spending, and a shift to government leadership and policies more friendly to the Soviet Union. Japan especially requires export markets for economic stability. If the huge American market were closed to these friendly nations, then trade with the Communist countries would become relatively more attractive to them, even though some unattractive negatives would be attached. In 1980 the total of U.S. exports to the Soviet Union was only about $200 million. The U.S.S.R. already buys ten times that much from western Europe and Japan.

Of course, we still could sell some of our surplus food to other nations. Except as limited by their counter barriers, we could sell them those technological items in which we remained superior; they might want to take advantage of the benefits those advanced products might yield them. Through these exports we would generate funds to buy some oil, minerals, and other needed goods from abroad. We would not completely fall apart, but the overall effect would be highly negative. Import restrictions, to bring the United States any real benefits, must be applied as exceptions, as we shall see more clearly later. It is far better to lead in technology than to invent defensive steps because we are lagging.

America's Slip in Technological Leadership

Unfortunately, we can hardly count on being substantially ahead of all other nations since the fact is that we have been slipping badly. In the past two decades we went from the long-held position of leading the world in annual productivity gains to the bottom of the list of all industrialized nations in this rate of growth. During the decade of the 1970s, productivity in U.S. manufacturing industries (measured by output per worker hour) increased by 28 percent, contrasted with an increase of over 100 percent in Japan and of about 60 percent in both France and West Germany. Because of their more rapid improvement rate, a half-dozen nations will pass us in absolute productivity in the next four or five years. We now already rank tenth in gross national product per person, behind every major non-Communist industrial country except the United Kingdom and Italy.

We have been losing out in world markets, as well as in our domestic one, in consumer electronics, cameras, automobiles, machinery that produces consumer goods (like sewing machines), and many heavy-machinery areas. The U.S. share of world trade for all high-technology products dropped from about 30 percent in the 1960s to 20 percent in the 1970s. Only in airplanes and spacecraft has our lead continued strong, and here western Europe and Japan are gearing up to give us serious competition. One in seven of all of our manufacturing jobs depends on exports. Loss of export volume is now a major cause of unemployment in the United States.

Studies have been made periodically of the sources of important technological breakthroughs. U.S. inventions and developments used to completely dominate such compendia in the past, but in the last several years less than half the listed items were of U.S. origin. The Patent Office keeps track of the patents it grants to foreigners. Ten to twenty years ago, such issues were only a bare fraction of the total, but foreign inventors have been obtaining about a third of recent American patents. Worse, patenting by U.S. inventors actually declined about 25 percent during the 1970s. Since western Europe, Japan, and the U.S.S.R. are together graduating far more engineers than we are, we can assume that our relative scoring of conspicuous firsts will descend further in the future. Of every 10,000 of its college graduates, the United States turns out only 70 engineers; the Japanese turn out 400. The U.S.S.R. now graduates over 300,000 engineers a year and Japan 75,000; fewer than 60,000 are being produced in the United States annually. The fraction of scientists and engineers in the U.S. labor force has declined steadily in the last two decades, while in this same period that fraction has doubled in both Japan and Germany. During the 1970s the number of Ph.D.s employed in the practice of physics in the United States dropped by 10 percent.

The annual number of Ph.D. graduates in engineering is an excellent indicator of future technological strength. With an increasingly technological society, this number should rise steadily. In the United States it dropped by a third during the 1970s. American universities are expected to award only about 2,500 doctoral degrees in engineering in 1983, compared with a figure of 3,300 ten years earlier. Foreign citizens now are obtaining over a third (moving rapidly to a half) of these doctorates in American universities, and a quarter of the junior engineering faculty in U.S. colleges are foreigners who received their advanced degrees here. It used to be that most foreign-born Ph.D.s from our universities became U.S. citizens. Now more are returning to their home nations to take part in the technological boom there. In the second half of the decade of the 1970s, U.S. corporations with operations abroad almost doubled the amount of R&D they conducted in foreign countries.

The number of engineers graduating from U.S. colleges and universities has not grown fast enough to keep up with our industry's increasing needs. Finding engineers and technicians is the most difficult problem facing U.S. high-technology companies. Some segments of our technological industry are growing at rates of more than 25 percent a year, and a shortage of engineers and scientists is all that keeps them from growing faster. It has been estimated that even in the present recession, 17,000 entry-level openings in industry for engineers are going unfilled. In small high-technology U.S. companies, foreign graduates of American universities often make up as much as half of the technical staff.[1]

There are 2,000 unfilled engineering teaching positions in U.S. universities. Personnel needs of American industry are expanding rapidly in certain critical fields, such as computers, while our universities are becoming short of professors and graduates in precisely these most promising areas. It may become necessary in the United States for engineering schools to depend substantially on visiting lecturers from industry, who will appear on college campuses part-time to introduce the students to the latest state of the art. Such a supplemental faculty would be advantageous but would make up only a part of the gap in meeting the nation's requirements. The situation building now certainly will lead to worse staff weaknesses in industry later.

The U.S. government has been concentrating on the immediate economic crisis and decreasing its contributions to the long-range goal of building university engineering strength to the higher plateau needed. The government, in fact, has recently been viewing engineering education as industry's concern. But industry leadership, busy running financially troubled companies in inflationary and recessionary times, has shown little interest in the problem, with the exception of a few companies in a few fields at a few universities. In both Japan and Germany the relationships among universities, industry, and government in expanding engineering education are much closer than those in the United States. In dollars, corporate support of universities has been rising slightly in recent years (from a level far too low). However, as a percentage of corporate pretax income (especially when properly corrected for inflation), the support has actually dropped. Various industry associations are now urging American technological industrial corporations to donate a mere 2 percent of their R&D budgets or 1 percent of their earnings before taxes to improve the nation's engineering schools.

---

[1] Yet jumping to the wrong conclusion that American companies must be developing a harmful dependence on a foreign labor pool (to the detriment of Americans erroneously thought standing ready to fill the openings), Congress is considering bills that would require foreign students to return to their native countries for at least two years after graduation before they can apply for the special visas issued to highly skilled workers the United States requires. Today aliens with advanced training are eligible to apply for such visas as soon as they graduate.

Judging by the poor results of similar past campaigns, only a small fraction of American companies will increase their contributions to these extremely modest figures.[2]

If nothing is changed, the decline of support by the government and the inadequate support by private industry of engineering and science in the nation's universities will ensure that the United States descends to a level of mediocrity and inferiority as an international technological competitor in future decades.

It has been estimated that half of the productivity increases in the United States in the last twenty-five years have been the direct result of technological innovation. An apparent confirmation of this estimate is the fact that research-intensive manufacturing industries increased their contribution to the nation's exports during the 1970s from some $5 billion to $30 billion per year, while in the same period, industries without strong technologically innovative tendencies, such as the auto and steel industries, declined greatly. By the late 1970s the latter industries represented most of the additions to the list of the nation's trade deficit categories.

The capital budgeting systems employed in many U.S. companies demand a relatively early payback (at most, five years), while new technology often requires five years or more for the R&D phases alone and still more time for product design and production implementation. Also, as indicated in an earlier chapter, the structure of rewards to U.S. managers of industry fails to encourage long-term risk taking, and this works to hold back the development of new technologies. Moreover, the effective tax rates steadily increased during the late 1960s and all of the 1970s. Since the greater the risk in innovating, the higher the after-tax return an investor expects, tax policy has discouraged innovation. Breakthroughs have become less likely.

The fraction of our total GNP going into research and development has been decreasing substantially, while in western Europe, Japan, and the Soviet Union, this same indicator has been rising. In the mid-1960s some 3 percent

---

[2] Innovative partnership arrangements in some special areas are now being worked out between industry and the universities. In such arrangements industry supports university research to obtain basic knowledge pertinent to future product plans. The universities make total public disclosure of the research they perform, a necessary policy for self-respecting universities, and most often own the patents that might result. The universities give the companies paying for the research a royalty-free license to use the patents to their advantage, while all other companies are expected to pay for that privilege. Such payments are added to the universities' research funds.

of the U.S. GNP was spent on research and development. By the end of the 1970s, it had declined to about two-thirds of this value, while Germany's figure rose by 40 percent and Japan's by 30. The U.S. ratio of purely civilian R&D expenditures to the GNP has been steadily below the figure for West Germany and Japan during the last decade. Since western Europe and Japan are producing many more engineers than we are, and are able to put roughly equal resource support behind each engineer, we can only expect our R&D ratio to lag further behind theirs in the future.

It is common to hear from those in government who have responsibility for arranging the government's funding of university research the claim that at least this figure has held up, even when inflation is accounted for. But this is not so if the inflation correction is properly made. The costs of modern research have inflated much more than the inflation of the economy as a whole, and general inflation indicators are misapplied if used on research funding figures to adjust for inflation. Anyway, even if pure research funding has stayed about even, that is hardly a satisfactory result, since the amount should instead be rising substantially every year, as it is doing for the nations against whom we are competing. Even worse, our government's steadily growing administrative rules have served to lower the true funding for basic research. They have added overhead burdens on the universities so that an increasingly large fraction of a grant goes not for the intended research but rather for lawyers, accountants, and experts on equal opportunity and other government regulations, and for sending scientists to Washington to explain their grant proposals in increasing detail.[3]

Over the last decade private investment in R&D as a percentage of U.S. industry's total revenues has gone down by about a third. A downward trend of many years has recently been reversed, but when properly corrected for the far-higher-than-average inflation in the costs of R&D, the expenditures in the past decade reflect a period of very substantially reduced emphasis on R&D. Our total research and development spending is still higher than any other single nation's. Unfortunately, we are not competing against any one nation, but rather the total of many aggressive ones. Also, since our products

---

[3] The United States is the only major nation that insists on funding government R&D on a fragmented, year-to-year basis; it has always rejected multiyear funding. Comes the summer and university scientists typically will still not know whether the next academic year's research programs will continue to be backed. Moreover they will not know whether to fire or hire their assistants, who as Ph.D. candidates need to know whether they will be able to go on with their education.

are losing out in the world market and our productivity is failing to increase, it is reasonable to look for and find significant differences between our focus on R&D effort and those of western Europe and Japan. Currently about $75 billion is being devoted annually to research and development in the United States by the combined action of government and private sources. Over 40 percent of these funds go to defense R&D, while in West Germany only about 12 percent of their R&D relates to military technology and in Japan the figure is only 2 percent. The bulk of military R&D effort does not lead directly to civilian products for the world market.

The competitive situation is more serious for the United States than these ratios alone suggest. Military R&D is concentrated on the most difficult of technical projects and so usurps a disproportionate amount of the country's technological resources beyond what dollar ratios alone would suggest. Designing a military computer may not call for a smarter computer engineer than does designing a civilian computer, but creating complex computer networks to provide command-control-communications-intelligence for the armed forces soaks up large numbers of hardware and software engineers. Similarly, a modern battle tank calls for more engineering time per dollar (or per pound) of total cost than the most expensive commercial truck. Again, the new Stealth bomber, by the time it enters the Air Force's inventory, will have taken up far more engineer-years than would have been necessary to develop the most modern passenger plane for the airlines. Finally, in military projects, much engineering time is used up in coordination efforts with the government. America's technological resources thus are severely overstressed in trying to accommodate both military and civilian requirements.

Still other factors are working against U.S. technological competitiveness. Starting with the 1970s, we have raised safety, health, and environmental protection to a high level. This has forced a heavy allocation of capital and scientific and engineering talent toward this objective. Even if we have been doing a good job of regulating industry operations so as to provide a sensible balance between safety improvement and cost (surely not always the case), every dollar spent to meet government regulations is a dollar that could have been invested in increasing productivity.

The Japanese have been saving from 20 to 25 percent of their income for the past thirty years, creating a large pool of capital to back their new industries. Our savings rate has averaged around 5 percent during that time. In the United States we do not have a savings policy that gives tax breaks to savers as against spenders. As a matter of fact, we still have taxes on interest income from savings and capital gains taxes on entirely productive investments, while Japan and Germany have none.[4]

The total funds we have been putting behind each worker to improve our manufacturing performance is below that of our principal international competitors. (Of course, our inflation has accentuated this difference; in the last ten years, Japanese interest rates have averaged half of ours, making investments in their industry more attractive.) In the middle 1960s, we ranked first in the world in capital investment per capita. By the middle 1970s we had slipped to sixth place. When the investment in really new facilities is pulled out of the total, it appears that over the last two decades the U.S. investment averaged out at below 3 percent of GNP, a ratio less than that of any other industrialized nation. The average age of a factory in the United States is twenty years, while in Japan and West Germany it is ten, so other nations' facilities are becoming more representative of the latest technology.

An example will illustrate our situation well. A few decades ago so-called materials-handling equipment, pioneered in the United States, used to be cited often as a prime innovation that had made a major contribution to the rise of productivity in American factories. The concepts and equipment implementations involved found their way only later into competitive foreign manufacturing. Today it is robotry that symbolizes the technologically up-to-date manufacturing operation. In this field, Japan is the acknowledged leader by a wide margin, and America is moving relatively slowly. In fact, leading American firms have begun to take out licenses on Japanese robot technology.

Efficiency Isn't Everything

It is not self-evident that attaining the highest efficiency in applying resources should be the end goal sought by a truly civilized society. Some of our slip from technological leadership has resulted from conscious decisions American society made in the last two or three decades to attain a different kind of advance, one more on the human level and one not counted by the conventional measure of productivity. For instance, it is important to attain

---

[4] The United Kingdom has an even higher capital gains tax base than ours; not amazingly, unlike Japan and Germany, they show even less investment in new technology as a fraction of GNP than we do.

reasonable safety in the workplace and in the use of manufactured products. Also, the effluents from the manufacture and utilization of products should not overly pollute the air, water, and land. If achieving safety, health, and environmental protection has required allocations of scarce expertise and capital, then so be it. It is hardly disputable that if the same resources had been devoted specifically to increasing productivity or inventing superior products, we would be more competitive today in the world market.

By the usual measures, a low productivity increase equates with a low increase in the standard of living. But if fewer goods produced is accompanied by less cancer, this adds up to a superior life in a different way. Similarly, fewer shoes produced per person per year, if accompanied by more time to walk barefoot on a clean beach, also may constitute a preferable existence. Efficiency isn't everything.

It is to the credit of the United States that we made a national commitment to curb prejudice in employment. It was social progress worthy of national pride when we became determined to bring millions of previously disadvantaged American citizens into the main employment stream. These moves required that we accept a substantial period of training to increase the future value of the nation's human resources. Such ambitions for improvements in the life, liberty, and pursuit of happiness in America did not happen to be compatible with seeking a maximum in productivity statistics that describe only the amount produced per worker-hour. This price we were willing to pay. We traded short-term productivity increase for long-term social gain possessing greater value in human terms.

There is more to our deliberate deemphasis of efficiency. By putting so much emphasis on military weapons systems, we may have been overly cautious or even hawkish—perhaps no one really can say for certain—but we know what we have been seeking. It has been to achieve national security, to ensure protection against potential enemies. If we have been balancing our priorities correctly, then what we have been buying with our defense expenditures and the consequent assignment of human and physical resources, is the preserving of our free society. Presumably, this has been worth more to us than what we might have captured in a rise in standard of living if those same resources had been applied differently.

To sacrifice efficiency in order to realize something we value even more is meritorious and certainly not mysterious. Anyway, with so large and diverse an amount of national activity, no way of organizing and decision making can bring us a perfect allocation of resources to our various goals. Indeed, one measure of value in our society, which we would all be loathe to see weakened, is the extent to which we are privileged to enjoy individual choices. Our lives are not overly organized by the government. If we insist on a strong commitment to liberty so that we can all follow our separate, uncoordinated, chosen ways, then waste and confusion sometimes will be more in evidence than neat, efficient, carefully planned cooperation. Similarly, if free enterprise is to be preserved, we must limit the extent of government direction of the nation's affairs. In practice, this often means that when the government is of necessity involved, it is as a regulator guarding against the most serious negatives that a totally uncontrolled society might produce. In turn, this frequently leads to a relationship of confrontation rather than of cooperation between government and the private sector.

In contrast with the looser, freer, and often adversarial American approach, government planning and industrial planning in most nations are done more in concert. But to emulate this practice of greater government control over citizens and companies would itself represent a lowered standard of living to many Americans, even if it were to cause our productivity to rise more rapidly.

To compete successfully in the world's technological Olympics, we need to design our own triangle of society-technology-liberty. We need to realize a configuration that will match our culture to our priorities in how we want our country to function, and yet still stimulate adequate technological advance. To accomplish this, however, we have to accept the fact that some powerful international factors are establishing the overall environment. What we set out to accomplish in our own way in America has to fit into the world as it is shaping up to be, not as we wish it were. Let us look at some of these worldwide influences.

The Contest between Free Trade and Protectionism-Nationalism

Two opposing approaches to international trade relations are active today. The first is a one-world, free-trade system. In this approach, natural resources, manufactured products, services, money, technology, management know-how, and even labor are interchanged freely, crossing borders without constraint. Each nation offers what it has or can most sensibly produce to others at market prices. It acquires from other nations in turn what they can best supply. In the second approach, the government of each nation tightly controls imports and exports and incorporates that control into its highly politicized, domestic problem-solving efforts. Every country has social and economic difficulties and a citizenry that looks to the government to improve things. The movement of assets internationally is bound to affect conditions within each nation. Thus it is a real-life impossibility for governments to keep their politics-picking hands off this flow.

The one-world, free-trade concept offers powerful economic benefits. When each entity concentrates on what it is best suited to do, those most fortunate in their possession of resources, skills, and developed infrastructures naturally will exploit such advantages. But if the most useful output that any nation, richly endowed or not, can provide is made available to others in return for whatever they can most readily contribute, then all nations will tend to be better off economically. Because of the significance of this potential economic gain, it is very difficult for governments to push the free-trade approach totally aside. It is without doubt a permanent factor in international activities. However, since internal problems are not going to disappear in any country, the protectionism-nationalism approach to international relations can also be expected to endure forever. For some activities and nations the free-trade approach will dominate. For others, the forces of protectionism-nationalism will be stronger. The commercial international affairs of the world will continue to exist partially in each of the two modes.

Technological advance militates toward a free-trade world. A nation that is strong in technology can add value to its natural physical resources. If its people are skilled in the application of science and technology, they will produce more efficiently whatever they are suited to making. Moreover, its engineers will conceive of new products and services that the world will want. When a nation's technology moves ahead, the nation achieves an advantageous negotiating position for world trade because all nations are anxious to acquire the leading technology.

This is why some of the OPEC countries are avidly determined to become technological nations in a short time. Not content to go on living on the income from selling their petroleum or to use their new wealth exclusively for financial investments in the more technological nations, they want instead to emulate those nations. They are trying to buy their way into the fraternity of technological countries. Crude oil is valuable in itself, but a newly fashionable, high-priority first preference with the countries that possess oil is not to be wholly dependent on the technology of others to extract and alter it into a more valuable product.

The poorer, less-developed countries (LDCs) also feel it imperative to elevate themselves technologically and thus become richer nations. Lacking the money to finance the changeover, they are seeking technological aid from the developed nations, because they consider it just as critical to their future as economic aid to stem famine or military aid to protect them against aggressive neighbors or internal rebels. They regard the more highly developed, prosperous technological nations as having a responsibility based on ethics and world statesmanship to provide generous assistance.

In all nations, developed or not, the one-world concept of free and open trade is especially fostered by exports and imports of technological products and the corollary transfer of raw materials. Crude oil may appear at first thought to be properly classified as nontechnological. However, the extracting, processing, and distributing of petroleum tie into essential industrial products which are technologically based. The world's petroleum problem is not one of supply alone but rather of the supply-demand relationship, with the demand dominated by technological applications.

Granted the new technological emphasis of world trade, the two classes of entities that most strongly influence the movement of assets among nations are the governments of the developed countries and the technological corporations that operate multinationally. Let us next discuss these major influences.

The Role of the Multinational Corporations

Multinational companies are key to international trade in technology because, except in some instances for the national governments themselves, they are the largest investors in technological advance and because, as they seek to maximize the financial returns on their investments, it is natural for them to view all the earth as their prospective market area. Other countries contain customers for whom a company's technological product may be as suitable as it is for customers in the home country. If a producer ignores the option of operating internationally, it will give sales away to more aggressive competitors who will then have the advantages of higher revenues and a broader insight for management and investment decisions. For most technological products, numerous applications exist. These change with time as the system of doing things is altered by technological advance or by changes in the society. A company that deals with a larger array of different customers in different countries will happen across more applications. International activity also broadens the technological base of a company for future product extensions because the company benefits from membership in the scientific and technological fraternities of all the countries where it operates.

Japan illustrates another dimension of the role of the multinational.

Almost devoid of natural energy and mineral resources and possessing exceedingly limited useful land area, Japan's economic strength depends on her developed technological resources—scientific and technological expertise, skilled labor, modern production facilities, supporting infrastructure, and a superb worldwide marketing system. Many of Japan's leading industrial corporations are fundamentally multinational because a large fraction of their products are designed specifically for the export market. Either Japan has to be superior in delivering technological products other nations choose to buy for reasons of quality and price, or it will fail to be a first-class nation.

Until recent years, because of the size and affluence of the American market, it did not seem vital to many U.S. technological companies that they operate in foreign countries as well as domestically, although the larger and well-established companies became international long ago. When a company has strong technological leads in some product areas, it often can merely export the products to increase its sales. If the product line possesses the attribute of high cost-effectiveness to the customer, that customer will be willing to pay the asking price even when higher than in the country of manufacture. In time, however, competitive technological advance being almost certain, each company has to consider manufacturing abroad to achieve lower costs there and survive against the eventual local competition.

Today the importance of the international market is appreciated even by the smaller U.S. technological corporations. Increased communications and decreased costs of transportation have brought vigorous competition in the U.S. market from overseas suppliers, but at the same time, opportunities have been enlarged for American manufacturers to expand profitably outside the United States. All technological corporations in the world now believe in moving quickly to attain a geographically widespread position and, through it, a superior return on investment.

In most developed countries, multinational corporations work closely with their governments, who support them in their building of favorable positions in world markets. A foreign government typically will give a locally owned company a favored status through its purchases of the company's output, as well as through financial assistance and tariff protection. Thus the company will have a better chance to build its strength for international competition by first having a captive domestic market. It is usually considered mandatory by large technological corporations based in other countries that they be active in the large American market. Foreign multinationals know they can expect assistance from their governments in taking the important step of coming into the United States.

The U.S. government and our own multinationals are much less likely to work together to reach national goals. First, U.S. goals are rarely clear, since essentially no attempt is made to articulate them. (For the purposes of this discussion, we do not count our public's almost certain preference for prosperity, domestic tranquility, and international peace as an adequate statement of national goals.) Second, government and industry leadership in the United States expend substantial effort in countering each other's desires and moves. Let us note some examples.

Both foreign countries and the United States have government-sponsored financing organizations intended to subsidize exports through the use of loans to world customers at interest rates below market rates. In other countries this idea is regarded as a sensible, established route toward economic growth. In the United States the concept is continually questioned and commonly described by influential critics in government and business as wrong in principle, wasteful, or unfair. Thus our Export-Import (Ex-Im) Bank is halfheartedly supported and regularly threatened with cutoff. The competitive disadvantage to American companies is easy to document.

The Boeing Company saw its share of the non-Communist world market for commercial aircraft drop from two-thirds in 1980 to one-half in 1981, owing to the cutback of availability of Ex-Im lower interest loans to its foreign customer prospects, while that of Europe's Airbus Industrie more than doubled, from 20 percent to over 40 percent, by heavy utilization of government financing below the market rate.[5] Two or three points of interest on financing can be more important to an airline ordering aircraft than a 5-percent fuel cost advantage. One year of squeeze on our Ex-Im loan budget lost Boeing some $4 billion of business. American nuclear equipment producers and telephone equipment companies have lost out to European manufacturers for similar competitive financing reasons.

In the United States, exporters lobby for the highest budget for the Ex-Im Bank to support their international sales, while others ask why the United States should underwrite improved financial performance for particular companies, benefiting their shareholders at the expense of other Americans who have to pay the regular, higher market rates of interest. Of course, there is a point here. Why not let the free market operate? Why subsidize American exports? To get at the answer, one must also ask why foreign countries do it. Do they understand something we don't?

Perhaps. As they see it, certain selected export sales that can only be

---

[5] Boeing's chief competitor is no longer McDonnell-Douglas, but rather Airbus Industrie, which has sold more than 500 jetliners to over 40 airlines.

gained by use of special credit subsidies will provide their citizenry with a long-term advantage, even though in the near term they have to absorb the cost of the interest rate difference. They believe it will provide a start for an industry that will grow, be very profitable later, and create needed jobs. In effect, through their government, the citizen-taxpayers are investing, making an immediate financial sacrifice with the hope of a good return later. In America we do not puzzle out the long-range alternatives on such issues with the thought of investing in the future U.S. economy. We simply allow the self-interest lobbyists for and against various government subsidies to fight each other for the available pot of immediate government funds. The winners are not necessarily those espousing the right view from the standpoint of the national interest but perhaps those that lobby most effectively.

Nearly $10 billion of Ex-Im Bank loans are outstanding in support of the U.S. commercial aircraft industry's past overseas sales. The nuclear industry is the second largest beneficiary of Ex-Im Bank loans, with around $8 billion on loan to date. The latter is the best example of confused government policy that does not help U.S. industry.

The nuclear electrical energy equipment industry at home is dying since no commercial nuclear reactors have been ordered in the United States since 1978 and others previously on order have been cancelled. It is the business the United States has landed overseas, in such countries as South Korea, Taiwan, and the Philippines, that has been keeping the industry alive. Meanwhile, the present administration advocates nuclear installations and exports. Its rhetoric is not only lenient with regard to accepting nuclear reactor risks to safety and the environment but emphatic about the idea that moving forward with nuclear energy deserves a high priority. At the same time, however, it has reduced funding to the Ex-Im Bank, which it labels as wrong in principle since it is a subsidy and an interference with the free market for money. Past nuclear loans have provided financing for construction of some fifty nuclear plants in foreign countries. Foreign competition is now much stronger, but our financing subsidies are much weaker. Thus terminal illness may be the fate of a U.S. industry that the administration wants to advance, but for which it is destroying an essential aspect of marketing. The real world, after all, is dominated not by a theoretically free market but rather by the actions and policies of foreign governments.

Let us take a different example of handicaps imposed by the U.S. government on U.S. business in international competition. A multinational corporation whose headquarters and ownership are in western Europe or Japan naturally cannot operate a plant in the United States in violation of American environmental protection laws. However, other governments do not tell their home-based corporations what they must do about environmental protection in foreign countries. In contrast, U.S. governmental regulations often follow our corporations, so that a U.S. corporation's foreign plant may be required to go beyond the regulatory standards the local government accepts, thereby giving a cost advantage to the U.S. corporation's competitors there.

Some important American laws written specifically to control the activities of American corporations are not applicable to foreign-based corporations operating in the United States. An American corporation seeking to raise funds anywhere in the world, for example, must do so within carefully prescribed bounds, but the American government does not bother itself about how foreign-based corporations operating in America create their needed financing if they do so overseas. To cite another example, foreign-owned companies in the United States were only recently required—and it took a U.S. Supreme Court decision to bring it about—to obey certain of our equal opportunity and civil rights laws in the hiring of their U.S.-based employees.

In the United States, large corporations are restricted by U.S. antitrust laws if they wish to share R&D expenses and technological data with leading competitors. The theory is that such pooling would give those companies an advantage over the rest of their competition, an unfair ganging up against those not included in the club. The government also has ruled that when companies employ commonly funded and owned technology, such action in itself works to lower competition and is illegal even if all competitors in the field are included. Furthermore, any one of a group of the strongest companies in a product area cannot expect to be allowed by the Department of Justice's antitrust division to merge with one of the others to form a combination more competitive in the world market.[6] The division would fear that a monopoly may result or that competition may be lessened in violation of our laws, even when it is a foreign company that is the really serious competition in our domestic U.S. market.

---

[6] To his credit, William Baxter, now head of the antitrust division, shortly after coming into that position, expressed thoughts that should have been government policy for decades. Specifically, he stated that during the 1960s, in its general hostility to mergers, the Supreme Court cooked up a variety of totally baseless theories about the harm they cause. The courts then penalized big companies for their size and interpreted the law as though it were intended to preserve the organization of industries in small units. "There never have been enacted into law any words from which one could conceivably derive such a policy," Mr. Baxter asserted.

Our government has forced large U.S. companies to share their proprietary data and license their inventions to all competitors to avoid a monopoly position, despite the seemingly opposite monopoly privileges a U.S. patent grants. This forced licensing by U.S. companies to competitors has been extended even to foreign-based competitive companies selling in the United States.

American laws on antitrust follow American corporations overseas. For instance, an American company operating in France may not be allowed to acquire one of its suppliers there, a French company, because the U.S. government might see the acquisition as a form of forbidden vertical integration. In contrast, a French company could buy that same supplier and then handicap its American competitor in France. The French government would either stay out of the way or actually encourage the acquisition in order to enhance the French company's competitive position against the American company. That French company could immediately proceed to exploit its newly augmented strength in the United States with no difficulties whatsoever with the U.S. government.

For decades, other nations' companies have formed joint export trading combines to market their products more efficiently in foreign countries. American competitive corporations have had to shy away from that kind of consortium because of our antitrust laws. Moreover, American banking firms, because of our banking laws, are not allowed to create or join as partners in such combines to provide needed financing, as in done in other countries. Congress is only now considering legislation to overcome these obstacles and permit American companies to join forces for stronger export marketing.

When Japan decides it wants to lead in semiconductor production, its industry and government work closely to achieve that goal by arranging a well-financed effort through as close a cooperative program in R&D as the leading firms choose to organize. The Japanese government dispenses funds to these semiconductor companies to broaden their development work. In contrast, leading semiconductor firms in the United States have to be exceedingly careful as to the degree and nature of their arrangement to pool funds for research and development or they will be deemed in violation of our antitrust laws. (We are speaking here of commercial projects. The DOD supports classified military R&D in which industry teams are permitted to organize joint efforts.)

For example, manufacturers and users of semiconductor chips in the United States recently launched a cooperative university-research program to help shore up the U.S. technological lead in integrated circuitry, which is being eroded by the advances of Japanese companies. To avoid running afoul of the antitrust laws, the program has to be open to any company wishing to join and has to be restricted to basic research, carefully avoiding product development. To satisfy the criterion of open membership, even a U.S. subsidiary of a Japanese company must be allowed to join, a requirement that goes directly counter to the program's objective.

The Japanese already have an equivalent program at a level more than twice that of the proposed U.S. project. The Japanese program is partially government-sponsored, and by no means is it open to Japanese-based subsidiaries of U.S. companies. The U.S. government, if it were seriously seeking the most advantageous world competitive position for American semiconductor companies, would be considering legislation to alter antitrust law so as to make possible the limiting of membership in the new U.S. program to U.S. companies. No interest by Congress in such a modification of the law has yet surfaced.

American automobile companies have also been handicapped by antitrust laws as to the degree of cooperation permitted to advance the technology basic to their product lines. They have even been prevented in the past from sharing R&D expenses and data for development of equipment for air pollution reduction. Unless U.S. law is altered, it would be unthinkable for America's "big three" auto companies to combine forces so as to compete more effectively against foreign car manufacturers.

So far in our discussion of the way multinational companies and governments are involved in the worldwide competition for manufacture and sale of technological products, we have avoided mention of an important factor, government control of the movement of technology across national borders. It is time now to consider that issue.

The Technology Transfer Problem

All countries now realize they should not rely entirely on their own resources in the technological race they consider vital to enter. Just as no nation, not even the United States or the Soviet Union, possesses enough of every critical physical resource, no nation has the best technological know-how to perform every task worth doing in that nation. Thus transfer of technology can pay off for the parties engaged in it, just as can the trading of any other item. The emergence of technology as an asset for transfer among nations and its rise to become the most important of all items of trade have also made technology a candidate for government control of its transfer.

During the decades immediately following World War II, the United States was the principal source of technological advance, and a steady transfer of our technology was made to Japan and the nations of western Europe. Recently a concern has arisen in America that competitive nations have been allowed too much benefit in the past from their easy access to our technological know-how. In allowing this mass export, we erred and now are paying for the mistake by the loss of competitive leadership. Whether to change our approach so as to deny others such ready acquisition of our technology has become a major political issue. Steps are now being considered by the government to control technology transfer more closely.

That we followed the wrong policies in technology transfer after World War II is debatable. True, had we held back technology from western Europe and Japan, their economic growth would have been more modest. What is not so certain is how the U.S. economy would have fared had we been more restrictive. Japan and western Europe would not have become so prosperous, and they would have been less able to purchase American products. All international trade would have been at a lower level. To have made sure the western Europeans and Japanese would learn little about our know-how would have required our government's setting up information communications controls that would have interfered with the open exchange of technological information that accomplished rapid progress here in America. We would have profited less from the typically free interchange in the United States of imaginative ideas, research results, and engineering data among our own professionals. Our nation's economic growth would have been curtailed, costs of production would have been higher, new products fewer, and our standard of living lower.

Under the most stringent attempts to prevent the flow of technology from the United States to other countries, our technological corporations would not have been able to operate internationally and employ the human and physical resources to be found in other countries to enhance the development of their U.S.-based technological endeavors. Admittedly, had highly restrictive import policies been adopted, foreign-based corporations could have been prevented from entering our domestic market to compete with American firms, and the latter would have had the U.S. market to themselves. But it would have been a smaller market. U.S. companies would have been the exclusive exploiter of U.S. technological advance, but less advance would have occurred.

All of these arguments, of course, are about the right course to have taken in the past. We cannot get back the technology we transferred, even if we knew for certain that we should never have allowed it to leave our shores.

Furthermore, the technological future will be quite different from the past, and it will be determined only partially by whatever technology transfer policies we now adopt. Most important is that we are no longer the overwhelmingly dominant source of technological advance. In a substantial part of the spectrum of all of science and technology, other nations now have the lead or are about to acquire it. As we have already noted, the total of Japan's and western Europe's engineers and scientists now exceeds ours, and their margin will grow rapidly. Moreover, with our particularly heavy emphasis on military applications, our talent available for international competitive endeavors in purely civilian areas is less than the rest of the non-Communist world is putting into the game.

If we were to erect barriers to keep others from getting access to our technology, we have to assume those walls would work immediately to stop technology's movement in the other direction as well. The other nations could elect to trade freely among themselves, and through active synergism could stimulate each other to technological advance to the fullest, while accepting our absence to whatever extent we ourselves bring about. In our self-enforced isolation, we would be limited to our own approaches. We would fall behind. If suddenly no technological products were allowed across our borders and such trade were to collapse, our economy would go into a depression. Output would fall and unemployment would rise. Our citizens would demand that the government act aggressively to handle the economic crisis, and we would end up paying for the attempt to keep our technology to ourselves with an economy heavily controlled by the government. More than ever, we would lack incentives to invest and invent, and our technological advance rate would deteriorate further.

Realistically, we must label broad or total bans as fanciful. However, might we not gain by narrow constraints on the export from the United States of certain technologies, and high tariffs on some competitive foreign products, the items in both categories carefully selected as exceptions? Granted, the selection process would not be easy, and any impediments to technology transfer we set up would create penalties for some portions of the economy, even as they might protect and enhance other parts. To begin with, who should devise the particulars for withholding our technology from the outside world? Surely not the multinational technological corporations, because they would simply act in ways they see as advantageous to themselves. It is not that they would disdain or ignore the overall national interest. Rather, it simply would be wrong to assume those corporations know what is in the national interest.

The federal government itself possesses only a limited understanding of that interest, skimpy expertise in the very technology whose export or import is under question, and slight knowledge of the world market for it and of the status of the competition. It has an equally inadequate knowledge in such questions as whether we really hold a lead and whether the U.S. economy will prosper most by openness, bans, or guarded secrecy. We have simply never set out to install this kind of know-how in our government.

American multinational corporations do not find it natural to think nationally. A profit generated in Hong Kong, Hamburg, or Bordeaux is as valuable as one arising out of operations in Chicago. (The Japanese seem to better understand their national interest.) But the U.S. government has not been of much help. I have had the experience more than once of seeking opinions at a high level in the U.S. government about whether one or another impending international deal for the company I represented should be deemed good for the nation. In each case I felt that the company's management was well qualified to judge the effects on the company and on the interests of the shareholders concerning specific uses of certain of our technology in foreign countries, but I assumed the management was far less competent to see how the overall U.S. national interest might be affected. However, I always found it difficult to get our government officials to express confident judgments. Sometimes they would refuse to offer any advice at all and readily admitted they really did not know if going ahead with the proposed project would be for or against the best interests of the United States.

Most technology transfer actually takes place through the activities of multinational corporations. As they go about their normal operations, they act to maximize growth and return on investment so as to satisfy their shareholders, their employees, the communities and governments where the plants operate, labor unions with whom they deal, financing firms, and other constituencies in whatever nations they are active. In working out strategies and compromises to please these separate and demanding interests, the managers of U.S. corporations, while not infallible, are more expert than anyone else since these duties constitute the essence of their jobs. For instance, if certain American inventions they own are made available to Japanese or European companies, it is almost certain that a proper quid pro quo, such as adequate royalty payments, always will be negotiated by U.S. multinationals. In truth, we have been selling, not giving away, our technology.

If the individual contracts for technology transfer to foreigners are negotiated skillfully by U.S. companies (admittedly, always in their perceived self-interest), then the combined compensation received may add up to a reasonably good deal for the American people as a whole. The French government may insist that an entity created in France and based on American technology be partially French-owned. Japan may require that the majority of ownership of a joint venture in Japan be in Japanese hands. Nevertheless, it would remain true that any new foreign-based entity, when it is finally agreed to and set up by an American corporation, is one that will presumably appear to that corporation's negotiators as advantageous. They are not in business to give anything away, and they are very clear about that.

Regularly, some in the United States invent the idea that the government should set up a special bureau to control such negotiations, to be sure that the best possible result will occur for our nation when technology is exported, or else to ban the transfer. But the personnel of any government agency we might create will lack understanding of the technological foundation for the negotiations, the business goals, and the market forces that determine how the activities being negotiated should be structured so as to be successful. Conceivably a government bureau, given veto power, might occasionally prevent a deal from being made by an incompetent American corporate management. But to catch an infrequent mistake would hardly justify bringing the government in and would only handicap timely, business-like negotiations.

Most large American technological companies have exploited their technology and made it available all over the world. Often such companies have been compared, concerning their record of domestic employment, with other U.S. companies that have remained exclusively domestic. These comparisons have always shown that the international companies, quite apart from their international growth, have increased their U.S. domestic employment more than the purely domestic companies. This seems to suggest that some U.S. companies should be encouraged to take their technology abroad as a route to general business success that would create more jobs for Americans here at home.

In deciding what degree of U.S. technology transfer should be permitted to other countries, our national interest is not a complete mystery. We would like to benefit from other countries' leads and maintain a generous list of leads of our own. When we are ahead, we want to be rewarded handsomely if we make our technology available to others. On the other hand, if a German or Japanese company happens to lead in some technology widely desired by the world, what is important to us is that we be able to enter into that technology adequately and soon. We want to have use of new and advantageous foreign-made products. We also want to participate in the employment opportunities that go with producing such products here, and this may require arranging a transfer of know-how to the United States.

In the best of worlds, all products and the technology to create them would move freely. The international output of technology in a totally free world would be determined by the talents, ideas, advanced positions, managerial competence, and capital available to the various implementers, each of whom, regardless of national base, would be engaged in working the world market, seeking the best return on investment. International companies would operate everywhere and even engage in joint ventures in various countries, combining their resources to achieve maximum economic benefits. But from the standpoint of the overall U.S. national interest, there is at least one unacceptable defect in relying on a free international market in the future.

We must assume that the future, like the past, will include hot and cold wars, economic and social instabilities, worldwide cartels against U.S. interests, and severe political world crises. During these supersensitive, dangerous periods, a minimum assured domestic output of certain products will be essential to U.S. security. We cannot allow ourselves to be totally dependent on other countries for that supply. For example, in a completely free world market we might find the United States totally out of the steel business, except for highly specialized steel-based alloys. We might not produce any miniature ball bearings. We might even find production of certain categories of semiconductor devices carried on solely abroad. In an emergency our government always could step in and take over any operation in the United States, even if it is foreign-owned, but such activities might not encompass all those regarded as critical. Thus we must establish a floor beneath our internal technological strength in key areas. We shall discuss this need further shortly. First, however, it will be helpful to examine some other facets of the international contest in which the U.S. government must interest itself.

U.S. Advantages Internationally

Since winning points in the international technology Olympics is important to us, we must try to understand our natural advantages and disadvantages and innovate to maximize the strengths and minimize the weaknesses. For the foreseeable future, the United States will remain the largest and most affluent singly integrated market for technological products. The Japanese market can be described as even more highly integrated, but it is smaller. The total western European market is roughly as large, but the individual nations there are often at odds with each other and choose to operate as a set of segmented markets. This means that if we were to provide advantages for U.S.

companies in our domestic market, it would go far towards ensuring their competitive success in world markets, whether in more speedily establishing a new product or in building the highest production volume of a more mature product.

As a single nation, we still hold the number 1 position in natural and overall technological resources, that is, in numbers of skilled people and facilities and in infrastructure that backs up technological activities with ready availability of raw materials, components, good transportation, and excellent communication. Also, we start out today with more leadership positions in various aspects of science and technology than any other country, even though specific countries surpass us in important fields.

We allow greater liberties to individuals and private organizations to engage in their chosen activities. Coordination by government is less obtrusive than in other countries. Free enterprise is broader and more active in the United States and makes more of a contribution through its inherent characteristics of motivation, competition, stimulation, and satisfaction. These are the U.S. advantages that we should be exploiting. The straightforward way of doing so would require that the game rules be those of a free market. Unfortunately, winning in the world arena of technological competition involves our often competing against government-private sector teams from other countries. We should next note some key aspects of these teaming practices.

The Role of Government in Other Countries

Other governments do things differently from ours, right or wrong. For one, they do more subsidizing of their technological industry and are willing to use their taxpayers' money to put together combinations of government and private investment in an effort to build strong starting positions in selected technological areas. For example, the French and the British governments combined forces and assembled their aircraft industry into a team to create the Concorde, the supersonic commercial aircraft. The same kind of teaming is behind the European Airbus program to compete in wide-body commercial jet liners. Other nations' governments are involved with their industry in setting competitive strategies for the world market and are more enthusiastic than is our government to become a full or part owner in technological industry.

They accept it as a national necessity that government should act to develop new industries and to keep specific industries alive—even when a positive return on investment for some of them appears forever doubtful. To illustrate, the French government came to the conclusion some years ago that France must have a strong computer industry. It saw no way for this to happen naturally against world competition in view of the relatively small size of the French market. So it stepped in and forced a merging of what might otherwise have been a group of mutually competitive French computer companies, none large enough to survive. It pushed into the combine not just the French companies but some French-located computer companies owned by foreigners (Americans). It then added government funding and gave the new French company the preferred position for all of its government business. The French did not believe it sensible to be totally dependent (as they were convinced they would otherwise inevitably become) on foreign-based companies to meet their essential computer requirements.

Most governments, other than the United States with its very considerable openness, place special controls on the activities of foreign multinational corporations in their midst. A company that is based elsewhere but that possesses an important technological lead can expect a warm welcome if it brings in its new technology, manufactures products based on it, and offers local jobs in that new field, thus creating a base for later technological advance originating in that country. The less-developed countries are particularly anxious to attract multinational corporations that will bring technology and create technological jobs. They often offer tax advantages and other subsidies. However, most countries have been greatly concerned for decades about their technological positions, and they present to technological corporations based elsewhere a double face. On the one hand, they are extremely anxious to pave the way for entry. On the other, they are suspicious and regard it as unsatisfactory when important technological activities are in the hands of companies from other nations. They continually connive with their own domestic corporations to gain advantages over companies based elsewhere, while simultaneously trying to attract such companies to their shores.

Even if the newcomers possess major technological leads, the extension of the welcome mat will often be accompanied by pressure to arrange some degree of ownership of the new enterprise by local entities. Sometimes the government itself is an eligible and zealous partner. Indeed, in countries other than the United States, such technological activities as the telephone system and the electric power generation and transmission systems are most often totally government-owned, and some of the important automotive, computer, aircraft, petroleum, and chemical companies have been taken over totally or partially by the government.

Other countries seem to be comfortable operating with a government-business plan worked out cooperatively. Their strategies are surely imperfect. They sometimes choose wrong projects.[7] Their plans are bound to be the result in part of short-term political pressures, perhaps to provide government rescue for an industry in trouble. But they start with the basic idea that having a national industrial strategy is wise—in fact, necessary—in today's world. The people fully expect their government and industry leaders to come together and figure out what ought to be done in the national interest.[8]

When other countries subsidize a field in order to get it started, or seek to penetrate the world market by offering artificially low prices or lower-than-market interest rates to finance sales, they are prepared to accept initial losses. On some of these projects they may never enjoy a net gain unless they can obtain a virtual world monopoly and through it force high prices later to recover their original investment. Of course, in the short term it actually is to the advantage of American citizens to obtain underpriced goods while other countries' citizens pay taxes to support that benefit to us. This contributes to creating a higher standard of living for us and a lower standard of living for them. But they are tolerant of a strategy jointly reached by government and industry to subsidize start-ups and strengthen their entries into the competitive world market, for they believe they most often will come out ahead in the long run.

That's the way things are done in foreign lands. Now let us examine how we handle similar problems and opportunities in the United States.

The Present Role of U.S. Government

U.S. politicians at every level of government are frequently quoted as enthusiastic supporters of free enterprise. They are outspokenly against the government's taking over and running industrial entities or otherwise

---

[7] The supersonic transport seems such an example, since there have been no orders for the Concorde from any of the world's commercial airlines except the government-supported airlines in France and Britain. It appears the Concorde will be a lifetime loser in a strictly business sense.

[8] For years, the Japanese, under their Provisional Measures to Promote Specific Industries, have allowed companies to take quick depreciation write-offs on robot installations and other advanced manufacturing technology investments. The law also allows low interest loans to purchase advanced equipment.

interfering with the free market. The steady communication over the years of these motto-type abstracts of government policy might yield the impression that the free market force is overwhelmingly the paramount influence in America in allocating resources and selecting the fields of technology to be emphasized. Financing investments, carrying out R&D and production, and maintaining the U.S. flag high in international technological commerce are all matters for the private sector. So, at least, the typical pronouncements of our government leaders would suggest.

Moreover, the argument is commonly made that the U.S. government cannot be of help and had best stay out of the international battle for technological superiority, that any government efforts to subsidize or set strategy would consist of mostly clumsy, delaying, and wasteful actions. U.S. government planning for America's internationally competitive technological future is considered by many influential Americans to be wrong from the start, because it would rest on the assumption that the government is competent to do such planning, when in reality it is not, and because the government next might seek to implement its plans, which would be even worse. Also, the claim is frequently heard that for the government to do strategic planning is especially an error because a better approach already exists. It is simply to allow the superior free-market system to work. To these critics, planning by government and industry in concert would be equally disastrous, even if imaginable.

But despite all the contrary impressions, the U.S. government is already very much in the act. The question is how much different and better we can make this government involvement—not whether it exists. Our government is heavily into every facet of the international battle for technological superiority. This has occurred without a strategy and without a plan, but instead through a series of separate, unintegrated, highly politically based actions (mainly on the part of the congressional rather than the executive branch), very little of which has ever had the deliberate objective of helping U.S. industry win any contest against foreign competition in the world market. Our pattern is that groups press for special government favors—a tax break here, a loan guarantee there, a price support somewhere else, a major project to be started or continued under government funding. Constituencies fight constituencies in Congress, trading and compromising with each other to obtain goverment assistance. The resulting, jumbled government decision making is hardly properly to be called overall national strategic planning.

Business leaders in the United States are at least as vehement and vocal as anyone else about keeping the government out of things and letting the free market work. But the principal executives spend a great portion of their time seeking government favors, contracts, subsidies, protective tariffs, special tax credits, or regulations that will help their activities and handicap competitors, both domestic and foreign.

The government continually dishes out subsidies. It legislates tax advantages for some endeavors and protects others with tariffs and quotas. There are government price supports for many products, and loans and loan guarantees by the government totaling hundreds of billions of dollars. The government aids some small businesses, including new technological companies, and the antitrust activities of the government have great impact on the size, merger possibilities, and product-line selection and expansion of big businesses. Japan engages in more deliberate, carefully thought-out, overt teaming of government and industry, but the U.S. government buys annually twice as big a fraction of the nation's GNP as does the Japanese government. (With military procurement counted in, our government buys over one-third of all the technological products produced here.) The government underwrites, chooses field of endeavor, and exerts administrative control over most of the basic research going on in universities. This, together with government-sponsored space, transport, energy, environmental, medical, and military R&D makes the government a deeply involved partner in roughly one-half of all technological activites in the nation.

Government regulations involving safety, health, and environmental protection influence greatly the capital budgets and allocation of resources for a large fraction of all of America's technological corporations. Licensing the use of the radio frequency spectrum, assigning geographical areas for telephone and video cable activities, and allocating communications satellite positions in orbit all permit the federal government to have a strong effect on the expansion of computer-communications services. The government is a nonsilent partner in the design of America's automobiles through its regulations on MPG, safety, and emission.

America has gotten itself into substantial federal ownership of industry, as exemplified by the TVA in electric power and Amtrak in passenger railroads. In these two examples, the reasons for involvement are quite different, showing that the scope of eligibility for government ownership cannot accurately be called narrow in America. Federal ownership within the electrical power generating industry arose from a perception that this was the only route to adequate growth in certain economically lagging regions. In contrast, the Amtrak program started when—partly as a result of bad government regulatory policies—an area of American industry was on the point of collapsing as a free-enterprise activity. It was considered necessary for the United States to continue to have a passenger-carrying railroad system, so the government assumed the ownership and operation. Since Amtrak operations do not show evidence of positive returns on investment, permanent subsidization seems necessary.

That the federal government accomplishes its subsidizing without a rational approach, a system of priorities, or a sense of long-term requirements is illustrated by its inadequate support of education in science and engineering, the basic source of future national technological strength. The government puts large funds into subsidizing the production of tobacco, spends further funds to prove that smoking causes cancer, but then absurdly eliminates the federal funds to update the laboratories of the nation's engineering schools, which must lay the foundation for the new jobs to which the tobacco industry's basically excess employment might eventually have to shift.

Of course, the foregoing statements about lack of planning do not fit our defense program. To be sure, the DOD can be criticized for insufficient long-term strategic planning and the White House for inadequate integration of military programs with other factors that make for national security. These matters we shall discuss further in the next chapter. However, by comparison with the government's on-and-off alternative energy programs, the out-of-date but continuing Clinch River nuclear breeder reactor project, the simplistic and oscillating approaches on the use of government land for petroleum and mineral exploration, and the decade of vagueness about the goals of the nation's space programs, U.S. military technology advance comes off as, relatively speaking, well planned.

Government planning specifically for the support of industry, either in the short term or long term, does not occur in the United States. Instead, when the U.S. government involves itself with an industrial operation, it is likely to be in an unprepared-for and frenzied way, in an envisaged emergency, perhaps to rescue a large U.S. corporation no longer able to compete successfully and faced with bankruptcy. Sometimes government loan guarantees are arranged for an ailing company. Such actions raise the prices to American consumers while maybe only delaying the time when an industry or company will have to be given up.

Backward or mismanaged companies are the ones most likely to run into economic difficulties and to push their workers into unemployment. When they do, political pressures are applied and pleas to be saved are made to both the legislative and executive branches of government. The approach by government may then be to go protectionist and try to keep out cheaper and superior foreign products (sometimes by blackmailing other countries into adopting "voluntary" quotas on their exports). In contrast, little political force ever arises to put U.S. government support behind accelerating starts in new fields of technology. An embryonic industry as yet employs few people, and it is not their jobs that are about to be lost. Moreover, old activities are entrenched activities, and the participants in them will be resistant to change even if the government were to interest itself in retraining and resettling workers so that they could be ready for employment in new technology.

In Japan, through joint government and private effort, deliberate emphasis is placed on building powerful positions in new high-technology growth industries. (In the past twenty years, growth in capital investment in Japan's semiconductor industry has been approximately twice the similar U.S. investment growth rate, while Japan's rate of growth in semiconductor production has been three times the U.S. production rate growth.) This is paralleled by an equally deliberate deemphasis by Japan of activities whose time is believed to be passing. In low-technology product areas, such as the production of shoes, consumer electronics products like radios and tape recorders, and routine kinds of steel, Japan's government-industry policy-making teams see the LDCs with lower labor rates as becoming more suitable places to produce those items for the world and they are investing in manufacturing facilities in LDC countries. In contrast, U.S. government programs do not exist to soften the inevitable phasing out of decaying industries.

A New Agency to Improve U.S. Competitiveness

Should the U.S. government be doing more? Is there not more the American private sector and our government should join to do? We are in a tough battle against other technological nations to put technology to the most innovative and efficient use and thus gain economic strength and growth. If other nations use teamwork by government and private sector to gain advantages in the contest, can we remain casual? How can we countenance adversary government-business relations in the United States when competitive countries are all taking nationalistic stands and their government and industry leaders are working together feverishly to outdo us? How can we argue that a free market offers the best results for our success in every competitive field when such a market clearly does not exist in key areas and cannot be forced by us into existence in view of the government-industry teaming in other nations?

In examining these questions, let us at the outset take it as a given that in the United States we should not abandon the concept that the free market remains the most efficient vehicle we possess for selecting the areas of technology worthy of allocation of resources. For many technological fields in which the United States has a good chance of being a world leader, although not all, the American version of the free-enterprise system will work well. It is also sensible to regard the American form of democracy—our tried and true, free-for-all, disorderly system, in which selfish constituency tries to outshout selfish constituency—as enduring for all time. Any attempt to depart radically from this approach, our ingrained way of life, will fail. In any case, the effort would be very unlikely to lead us soon to great breakthroughs in strategic planning.

Moreover, our pattern cannot be changed in the near future to one where close government-industry cooperation (a la Japan) will become the favored way to arrange for American world domination of important technological fields. Neither bolder leadership by industry nor the bursting of bloated bureaucracies in government will result by our trying to create a consensus-controlled committee of industry and government to run technological advance and try to beat out international competition. However, even after granting all this, some innovations—exceptions to the rule of no governmental strategic planning and no government-industry teaming—could make a big difference in the U.S. position.

Consider first a certain small-step start. Suppose the U.S. government, assisted by a group of expert advisers from industry, were to develop and maintain an up-to-date listing of potentially dwindling industries and of new ones expected to grow rapidly.[9] With regard to the first group, other countries might well win the eventual competition, whereas we might stand an excellent chance of being the world leader in the second group if the right things were made to happen. The backward group would consist of those industries where other nations' lower labor costs might prove dominant or where the technology is old and straightforward and at a level of sophistication now attainable by many competitive nations. The growth group would include those industries where the strong characteristics of America—advanced technology, specialized skilled labor, a large domestic market, and

---

[9] The Business Roundtable, an organization of chief executives whose 200 members are drawn from America's leading corporations, already is active in pondering leading business-government problems. It might be called upon to great advantage for analyses and ideas, just as the government, in matters of science and technology, draws for counsel on the national academies of sciences and engineering.

high capital investment—would be judged to be paramount factors for success. Also, in this second group substantial synergism might exist between military and civilian applications of the same technology.

As an example of the first group, consider the production of the very smallest of passenger cars. Here the domestic situation in some nations might compel them to take great interest in the most compact of low-powered vehicles and they might therefore specialize in producing them. Those countries might have certain advantages over us in producing such automobiles in higher volume and at lower cost. As an example of the second, technologically more sophisticated group, look at commercial airliners. Here our military programs in advanced aircraft should be an important factor giving us a head start. Moreover, our large domestic airline market, based on our wide geographical span, a large number of frequent travelers, and a broad technological infrastructure that supports our aircraft R&D and production, should set us well above every other nation or even most imaginable combines of other nations.

Let it be clear that we would not want the government to take action to impede the progress or injure the health of an activity merely because it is listed by the government as potentially fading. Who knows what unforeseen innovations in management, engineering, or marketing approaches might keep that U.S. entity a strong world competitor after all, despite the apparent trends? However, the U.S. government cannot, through subsidies, convert a noncompetitive American company into a competitive one, so drawing up the list might alert the government to be careful about involving itself in hopeless rescue attempts when the listed candidates for possible extinction really deserve to be in that category. The government might better engage in trying to ease the transition period of expected decay.

Is there not sensible action the government could take to encourage the timely retraining and relocating of employees, thus easing the burden on people who are displaced by technological advance? Again, first reliance on the free market (for labor) is important. However, those employed in industries that will not require their future services get little and late warning or help in preparation for change through the workings of the free-market system alone. Unemployed or soon-to-be-released workers cannot afford to pick up their families and move about the country, or enroll in costly retraining programs, acting merely on rumors or hunches about possible reemployment in coming technological fields or new geographic areas. Here the government might be an objective fact finder with citizens' interests in mind and be able to decrease the pains and accelerate the pace of necessary changeover.

The government might offer inducements to encourage the listed new-growth industries, those needing new facilities and added employees to carry out their expected expansions, to plan their buildups in those areas of the nation where existing employment levels are expected to drop. The inducements might include a sharing of expense to retrain workers already resident in the area so they can fill the new openings.

Let us turn next to the problem of our antitrust policy as it affects the U.S. status in the international competitive battle. The Department of Justice has a division with the mission of searching out actions by corporations believed to violate the antitrust laws. This group naturally has neither the urge nor the duty to invite competitive companies to engage in a level of cooperation that will get them into serious trouble with that very government department. However, suppose a new government unit were created, perhaps in the Department of Commerce—let us call it, say, the World Commerce Agency—and assigned the function of pressing for cooperative effort among competitive American companies with the goal of beating out foreign competition. (This new World Commerce Agency might also be the very government unit we just proposed that draws up the lists of diminishing and growing industries and assists in funding the retraining of personnel.) When it sees emerging growth industries that appear to have outstanding possiblities for new employment opportunities and increased exports, the agency would promote action among U.S. companies to organize common programs to help their position vis-a-vis foreign-based companies.

On occasion, as a result of its consultation with industry to plan competitive success internationally, the World Commerce Agency might press for mergers among American companies to create stronger entities. Acting in this way, it would move almost counter to the Department of Justice's antitrust division, so the two government agencies, if we did nothing else, would soon engage in mortal combat. The legislation creating the new agency would have to be accompanied by amendments to the present antitrust laws to incorporate a different, more modern meaning to the idea of competition, one that brings the international arena into proper focus.

We must advance beyond the existing legal concepts of monopoly and competition because in the main they are sensible only for purely domestic situations. Specifically, the new law should allow that whenever seriously threatening foreign-based competition exists, American companies may combine forces to some carefully prescribed (previously illegal) degree. It may well turn out that merging two or more U.S. corporations could create a near monopoly in a theoretically isolated America. However, we would not have to fear the ill effects of a domestic-based joint effort by American companies if there is certain and real competition in the United States from a large foreign combine, especially one where other governments are teamed with and are financing their corporations. No monopoly would then result, the competition within the American market would be ample, and U.S. companies would be able to compete more satisfactorily in the world market.

The new law should particularly encourage the development of growth industries by permitting joint activities by American competitors whenever technological superiority by the United States is believed important to us and possible, and the competitors are seen to be government-industry teams from other nations. Such foreign government-backed competition makes appropriate joint defensive measures in America sensible. The proposed World Commerce Agency should be chartered to judge such joint activities and approve them if they meet the stated criteria.

If we believe the free-enterprise approach brings advantages, then we must recognize that some technological activities—new commercial airliners, certain alternative approaches to petroleum-based energy, some space applications, for instance—can be attempted only by very large and heavily financed corporations. To emerge as a winner in this class of key international competition, American needs a substantial number of supersized companies. If we wish to avoid having no choice but to rely on the alternative of highly politicized government overinvolvement in such necessary but huge and speculative project areas, large corporations must sometimes be allowed to cooperate closely, even forming syndicates so that they can share effort and risk. This especially would be a superior approach for programs requiring a long period and an enormous investment before there could possibly be a turnaround to a profit situation. The procedure would call for those large companies wishing to form a joint venture to apply to the proposed World Commerce Agency for an antitrust immunity ruling that it would be empowered to issue, based on the particulars of world competition and other detailed facts associated with each individual project.

The new agency's role in the international civilian scientific and technological battle should include still another important responsibility. Since the military R&D effort is so large a fraction of the total of all science and engineering in the United States, and since this area is totally under the control of the government rather than the free market, considerable government planning would appear to be required and justified to ensure that developments in military science and technology were exploited to the maximum to advance the U.S. competitive position in civilian technology products.

Admittedly, taking advantage of the relationship between the two classes of technology is part of the management strategy of those American corporations engaged in both military and civilian fields. Actually, virtually all leading technological corporations in the United States perform at least some military work, even if they are primarily civilian products companies, and the opposite—military contractors engaging in selected civilian product endeavors—also exist. Since a corporation has limited resources, it sometimes decides whether to respond to the opportunity for a military technology contract partly on the basis of the potential benefits of such an effort to its civilian business. In some fields, the aggregate effort of many technological corporations, each going about its separate optimization of its military-civilian product mix, may come close to representing what is good for the nation as a whole. But in general this is not going to occur automatically. The government should be active in scheming with industry on how to realize the fullest exploitation of U.S. military R&D in the development of civilian products for international markets because the government alone has access to the indispensable knowledge and powers needed for national success in winning the competition for such markets.

If a new government unit is to make substantive contributions rather than merely to add bureaucratic confusion, we must first of all recognize that while our military technology has enormous potential for enhancing our civilian technology position, the need for restricting classified information cannot be bypassed. A difficult sorting job has to be done, and it can be accomplished only with the cooperation of both industry and the Department of Defense. Together they would select specific technologies that could be removed from the classified category because their declassification would do little or no harm to our military strength while their utilization in civilian applications would bring very high rewards in economic strength, both strengths being important to national security.

This is not a mission properly placed entirely with the Department of Defense. The DOD must concentrate on defense matters. It should not be expected to take on a prime responsibility for the nation's winning of civilian product competitions.

The role should be assigned somewhere in the government for chairing this selection and technology transfer process, working out the security versus economic priorities in concert with all government and industry groups concerned, and then seeking to exploit the potential in accordance with a deliberate strategy. No government unit today has a charter to handle this function of integration and coordination. It is a major disadvantage in winning worldwide superiority in technology that we do not have an organizational structure, a pattern of communication and decision making, that tunes all our technology strings to play a single, harmonized national strategy melody. If this situation is to be improved, the White House inevitably would have to participate in the process on a policy level, but the staffing of this function for detailed execution might also be placed with the proposed World Commerce Agency.

Let us shift to another area of international technological conflict where government planning is essential. The Defense Department already engages in identifying those fields of technological industrial endeavor where some minimum domestic American activity is important. Moreover, the DOD recognizes that some government subsidy may be needed if American companies in any of these special fields are losing out in world competition. The World Commerce Agency, concerned as it would be with all government efforts to strengthen the competitive standing of U.S. industry in the international civilian market, could play an important role here. The DOD alone must originate the list of products and specify the quantities and production timing needed for national security. Also, the DOD should be the main source of subsidy funding, since this should center around price guarantees and stockpiling for the Defense Department's later use. But the DOD's mission is defense, not international competitiveness. Hence the value of a World Commerce Agency in this situation.

Inadequate U.S. domestic production of critical items usually occurs because our industries in those fields have lost their competitive positions. In deciding when and how to subsidize a minimum domestic output in some area, the U.S. government first should learn why our industry has lost out against its foreign rivals. If that industry segment is about to be aided in the interest of national security, then some account should be taken of the incidental impact of the contemplated government help on the competitive health of that industry unit in the world's civilian market. If the industry is of borderline profitability, the guarantee of government purchases to fill security requirements at a realistic, if not generous, price may contribute sufficient profitable volume to improve the industry's chances of staying in business as a civilian sales competitor. The output of the products both to meet required defense purposes and to fill a civilian market demand should be combined in estimating the total U.S. industrial activity in the field. In turn, such a forecast will lead to a better decision about the nature and amount of government subsidy that is sensible, considering that two quite different markets make up that U.S. industry segment.

Will the World Commerce Agency help to make America more competitive in international technology battles, or will it hamper and confuse industry, waste government funds, and simply be another government bureau interfering with the superior free market? The answer lies in how wisely the charter of the agency is articulated and the way is it allowed to operate. First and last, the agency should act strictly on an exception basis, that is, in only a selected number of problem and opportunity areas, those that potentially possess outstanding leverage in determining the nation's competitive position and where free enterprise alone will not enable the United States to win the world contest.[10]

For success in these singular missions the agency will require a charter with clearly defined responsibilities and authority, a proper staff, and adequate funding. We have described some agency functions that would appear to be sensible. While these do not necessarily constitute the optimum or complete charter for the agency, let us summarize them.

1. **Antitrust.** The present laws should be modified to allow for joint effort and mergers among American companies, large and small, to create strong world competitors and allow for the formation of joint world trade companies. The new laws should give the World Commerce Agency the authority to act favorably on requests by U.S. companies for antitrust immunity rulings when (1) foreign competition in the U.S. market is severe enough to ensure no domestic monopoly could result from a favorable ruling, (2) the prime competition is a foreign government-industry team, or (3) the project involves so great a risk that a syndicate of companies to share it is necessary.

2. **Subsidization of essential industries.** The agency should be budgeted to assist private industry in those fields where, without such government assistance, the industry might cease to exist in the United States, but where a minimum capability for security purposes is deemed essential. The Defense Department should continue to determine requirements and cover most of the needed funding through stockpiling and guaranteed pricing. The agency should

---

[10] It is reasonable to ask who will be the judge of this. One answer is that the agency will proceed as its leadership decides, probably after generating criticism from within the industry that it is getting into too many areas or into the wrong areas and neglecting others. This criticism (and the resulting arguments in Congress, in the media, and between the agency and industry) itself involves the kind of joint strategic effort between the government and the private sector we should favor and seek to expand.

develop judgments on the nature and amount of subsidy needed in view of international civilian competition.

3. **Transitional support.** The agency should identify fields of industrial endeavor expected to phase out in the future as well as anticipated new high-growth fields. It should ease and encourage industry changeover and aid in relocating and retraining personnel. It should help fund the new industry's personnel training in return for that industry's willingness to locate in regions where older industry is phasing out.

4. **Export financing.** The agency should advise the Ex-Im Bank on fields of technological endeavor in which it is important for U.S. companies to be successful in competing against foreign companies. The agency should have some interest-rate subsidization funds to transfer to the Ex-Im Bank as a supplement (covering the incremental interest payments) to make U.S. companies more competitive whenever the key competition comes from foreign government-industry teaming.

5. **Strategic planning.** If the agency is competently involved in activities such as those already listed, it will undoubtedly attract the nation's technological industry into cooperative planning because of the potential aid it will offer in winning world markets. Thus, on an exception basis, focusing on the areas where the United States is today being bested by the government-industry teaming of foreign countries, effective government-industry strategic planning would begin to occur in the United States.

U.S. Statesmanship in International Competition

The government action needed to support American industry in the international competitive battle includes more than the specifics just described. Long-term stability and consistency of government policy is paramount. Our competitive technological stature is affected greatly by how the government fights inflation, high interest rates, and unemployment. Equally pertinent are the government's budgeting for basic scientific research in the universities and for military R&D. Finally, the government's approach to international diplomacy, export-import financing, technology transfer barriers, and environmental protection are all extremely influential.

Our federal government's record in handling these policy issues discloses far too much variation, ambiguity, indecisiveness, and superficiality of attack. Without clear, sound, and steadily maintained government policies, the U.S. private sector cannot be healthy enough to compete strongly and it will invest only modestly, with little confidence. Unacceptably high waste results when free-enterprise investments in new technology must continually adjust to oscillating, overly political, and confused government policies. We cannot pass a constitutional amendment that requires of Congress and the executive branch that their policies be more vigorously and deeply pondered and stable. But our citizenry has begun to perceive that longer-range strategy is important to all of us, and we can hope that this understanding gradually will come to influence government and industry leadership. Then more long-range, and less purely short-range, policy formation will result.

In the foregoing discussion, we have raised the question of the need for some joint government-industry planning if the United States is to improve its performance in the international contest for technological superiority. This cooperative effort must be categorized as more nationalistic in approach than is true of our present pattern in world commerce, even though it is action taken to counter the nationalistic actions of other countries. At the least, it certainly does not qualify readily as a step to broaden the practice of open, free-market world trade. But unfettered movement of technology remains not only an ideal, but a practical route for much of the technological activity of the world. If, to protect its competitive position, the United States here and there should become more nationalistic, would this bias all nations toward further protectionist policies? Would the partial abandonment by the United States of the goal of openness and nonprotectionism in our own large domestic market doom the idea of a free market for the fraternity of non-Communist countries?

That depends on how we do it. When the U.S. government backs U.S. companies' export sales in world competition by offering to match low interest loans by other governments to potential customers, we should make clear that we prefer that every country rule out such measures. Aggressively, we should propose to reduce government subsidizing through step-by-step agreements among all nations, which will in time ban such subsidies altogether. Similarly, we should lead in pushing mutual policies that will open domestic markets and procurement by all governments from the lowest bidder of quality. Our domestic market is so important to other nations that if we impose any restrictions at all on entry by foreign companies, other governments have to take such action very seriously. Clear, steady indication from us that we are ready and that we highly desire to abolish all such restrictions in return for equally improved freeing of their markets may cause progress toward a one-world open market. However, we cannot offer to ease our barriers in an effort to get others to follow suit if we have no barriers to ease.

The United States is the only nation that can exercise the statesmanship needed to move the non-Communist part of the world toward common practices in which technological advance benefits everyone. If America, both in its governmental and industrial leadership, performs better in the long-range managing of its technological endeavors, then its influence in the world will be enhanced. On the other hand, if our position continues to slip, then not only will our living standards suffer, but our country will also become incapable of assuming the role of statesman-leader in utilization of technology.

The triangle of society-technology-liberty for the world is not drawn properly if it describes the battle for technological superiority as a totally penalizing contest between nations. Competition should be sought because it is an important ingredient for stimulating maximum performance, but no nation should regard any other nation as a rival whose victory would spell its own economic doom. That would lead to a totally undesirable triangle, where the liberty focus signals each country's economic insecurity, the society focus issues severely defensive controls, and the advances and benefits from the technology focus are available to none. The non-Communist world must aspire to a triangle which recognizes that every nation gains from technology when the technology is offered to all. Know-how, materials, products, and money should move freely among nations. Constraints should be rare exceptions, not common rules.

Chapter Eight National Security in a High-Technology Age

Can anything be more important to our nation than what we simply label national security? Is there some other benefit that technological advance might provide us which should outrank the preserving of our freedom? If we face the threat that our lives might come under the control of unfriendly nations whose objectives and standards do not match ours, and if innovative and sound use of science and technology can help remove that threat or at least hold it to a tolerable level, does that use not merit the highest of priorities?

We surely have been acting as if the threat were real. We are planning to spend $1.5 trillion in the next five years to make ourselves stronger militarily. But how do science and technology fit into the program we seem to have espoused to make the nation more secure? If total destruction from Soviet nuclear bombs detonating over America is the big danger, then our earlier discussion suggests that the most sensible action to reduce such a peril is hardly to be categorized as wholly technological. It has to do more with the skill of the nation's leadership in international negotiations, because what we need is to arrange a workable agreement with the Soviet Union to greatly reduce nuclear weapons.

But even if this arms-control approach is pushed hard and works surprisingly well, it will cover only the most awesome of the identified security risks. Numerous others will remain or arise. The world will continue to be plagued with serious social and economic problems. Deep hatreds and misunderstandings seem to exist almost everywhere, and military force is being resorted to every day in many places. In looking out for our own defense, we will continue to exert a measure of world leadership, provide aid to some groups, and apply pressures on others. Meanwhile, the Soviet leaders will be busy seeking to protect what they perceive as their security. The result of all these circumstances and activities is that the world is approaching the trillion-dollar level of annual arms expenditures, counting everything from research and development of technological weapons to equipping and maintaining armed forces.

The advent of the atom bomb certainly has shown that scientific and technological breakthroughs can make an enormous difference in the power of individual nations to attain control over others. Even without further advances in military technology of such cataclysmic character as nuclear weapons, the continuous worldwide pursuit of scientific and technological advance has the dangerous possibility of uncovering new concepts with potential applications just as perilous, and the world will always possess enough severe social-economic problems to provide an environment in which such new technology will be nurtured and perhaps employed. The society-technology-liberty triangle will remain key in the geometry of our lives.

So it is important to look at the role of science and technology in providing for U.S. national security. What should that role be? Are we using science and technology to bolster security in proper balance with everything else that relates to security?

Technology—Necessary but Not Sufficient

To be secure, the United States needs many things: economic strength, social stability, high morale and patriotism, an understanding of potential enemies, skill in formulating foreign policy and negotiating with other nations, a broad industrial infrastructure, assured availability of resources for the anticipated duration of possible wars, an effective organization for setting security strategies, and adequate military forces. A solid defense posture requires integrating and balancing these diverse items, a difficult but necessary systems task. This list of mandatory requirements for national security is long, and military force is only one factor. That one component, adequate military strength, must be well matched to the other components of security, and it has a rather varied set of subrequirements of its own. One of these is weaponry, including not only those weapons based on recent and complex technology but also military hardware that is more mundane, simpler, less technological. To guarantee a sufficient quality and quantity of the high-technology weapons alone, the United States needs science and engineering skills in depth. Over the long term, this requires a continuing national program that plants the seeds for and cultivates the expert human resources behind technological advance and makes sure that an array of technological projects specifically geared to military needs are constantly being started and carried forward.

Technology is neither the basic cause nor cure of wars. In the defects of homo sapiens, not in our tools, are lodged the real origin of war-peace issues. But a modern nation with the goal of enforcing its wishes on the world militarily must take the high-technology road to reach it. Conversely, a country that lets its position in science and technology erode will endanger its security. That the United States must not allow itself to be a poor second in advanced technology to any potential enemy is hardly a subject for debate. Technological strength is a necessary condition for a credible defense posture. But it is far from a sufficient one.

In relating the scientific and technological efforts of the United States to our overall national security, we must keep in mind that what happens on the scientific and technological front will affect many of the other requirements for security just listed, such as social stability. Our economy is very sensitive to the size and nature of our overall efforts in science and technology. Thus we must not place emphasis alone on the technology leading to weapons systems and inadvertently deny ourselves the availability of sufficient talent and resources to achieve, through technological advance, the economic strength and related social tranquility required for national security. The two essentials, civilian economic strength and military power, compete in their common dependency on technology. Such competition would not be a matter of concern if we possessed technological resources to spare and enjoyed clear superiority over other nations in military weapon systems as well as in civilian technological products. Unfortunately, broad and overwhelming preeminence by the United States, as we have already observed, is not realistically to be expected any longer.

It would be fooling ourselves to set down the goal that we be superior to the Soviet Union in each and every dimension of military technology. The priority aim of the U.S.S.R. is to be at least as strong in weapons systems as we are, and the CIA estimates their military R&D effort to be twice as large as ours. Neither nation has infinite resources and both must meet requirements other than military ones, but the Russians are in a stronger position than we to force their citizenry to accept a stagnant economy and a low standard of living. Hence they can assign a larger fraction of their total resources to the military. On the other hand, we have a substantially higher per capita production, and our strong civilian R&D and overall industrial activity indirectly provide very substantial enhancement of our military efforts. While they can allot more bodies and tons of equipment to their military, we are capable of putting more technological effectiveness behind each individual in the force than they can.

It would be useless to attempt an accurate prediction of the relative future standings of the United States and the U.S.S.R. in military technology. As scientific and technological advances occur over the next ten or twenty years, an intelligent guess is simply that the competitive score will be close. The important thing to remember is that since we possess limited resources, U.S. superiority in critical areas will be achieved most often if we can be especially wise in picking which technological areas and projects to emphasize.

The Need for Basic Research and Development

Our ability to select new technology areas for military applications is based on the depth of our understanding of the underlying science. A broad research program thus is mandatory. First of all, it would provide us with better odds on being the nation that leads in uncovering new scientific principles that may have important military applications. Also, it would enable us to respond quickly, should others jump into the lead. Had Hitler's technical experts alone possessed a comprehension of the atom sufficient to develop nuclear weapons while the rest of the world was too backward in science and engineering to be able to match this capability quickly, Hitler surely could have used the atom bomb to control the earth. If, after World

War II, the citizens and government of the United States had been Nazi-like, our monopoly of nuclear weapons would have yielded us complete domination over all other nations.

Unfortunately, atom bombs cannot be considered the end of the stream of scientific discoveries and technological developments which from time to time provide a terrifying advantage to that nation originating the breakthrough and possessing the capability to follow it up. For instance, scientists have recently broken the genetic code. The techniques for creating new living substances are now being detailed. Who knows what new and powerful weapons systems might stem from crashing the frontiers of biological science? Or, as another example, consider that although nuclear explosions have multiplied by a million times the energy that can be released from a given weight of matter through conventional chemical explosions, the laws of physics suggest that vastly more energy can be released through an engineered obliteration of mass. It is not yet known how to achieve what is now a theoretical possibility, but a sudden unfolding of the technique could conceivably result in havoc for the world if the wrong people were the first to come upon the knowledge. As a final illustration, if a breakthrough permitted an essentially perfect, yet practically attainable, means to destroy enemy nuclear warheads directed at us before they could be detonated, it would completely alter the role of nuclear weaponry, eliminating the threat and the need for nuclear deterrents.

Still another reason for a vigorous research program to back up U.S. security is the need to bridge the technological and the nontechnological aspects of security. We need to be able to confer competently with other nations on war-peace issues. In negotiating nuclear arms reduction, we have to discuss eliminating existing weapons even as advanced ones are in development. We are interested in nuclear reactors for the generation of electric power, and we want to maintain a leading position as exporters of nuclear reactor equipment. On the other hand, proliferation of nuclear weapons is one of the world's most serious threats. We have had terrible trouble arranging nuclear arms reduction by the two superpowers, so we can have no enthusiasm for seeing more nations come into possession of nuclear bombs. Already, some fifteen or twenty additional countries could assemble the necessary expertise, materials, and equipment to produce a small but frightening number of nuclear weapons. They could obtain the critical fissionable material for bombs by extracting plutonium created in the fuel rods of conventional nuclear reactors as they generate electric power.

Thus, handling the problem of nuclear proliferation includes knowing what kinds of equipment and know-how should be held back from transfer by the advanced nuclear nations to the others. We need to know how present and future bomb designs and their required fissile materials relate to the designs and radioactive products of nuclear reactors that generate electricity, and whether the phenomena involved in turning nuclear reactor products into bomb material can be monitored from a distance, say, from a satellite. Expertise in the continually advancing science and engineering of nuclear phenomena is required for success in peacekeeping diplomacy.

We wish to lead in developing computers and applying them to the U.S. civilian economy, and we want to prosper as a computer supplier to the world. However, computer science is basic to military communications, command, control, reconnaissance, and intelligence. Thus it is vital to stay ahead of the Soviet Union in this field, which means it is important to deny them ready access to our computer advances. These desires, on the one hand, to distribute our computers for commercial gain domestically and internationally, and on the other, to prevent the transfer of the technology to the Soviet Union, are likely to be in conflict. Balancing the opposing requirements is an international policy problem involving numerous security, foreign policy, and economic facets; it cannot be correctly described as a matter of science and technology alone. Without an adequately broad-based and continually expanding body of computer science, our policy deliberations for the present and future might make little sense. What kind of information useful to their military might the Russians glean from acquiring our civilian computer hardware or the basic information inherent in its design? What new ideas are in the offing that should be kept secret because of special leverage in filling military requirements?

The horizons are limitless for further scientific and technological breakthroughs that can come to dominate national security concerns. Unfortunately, while unforeseeable quantum leaps in science and engineering by potential enemies could be catastrophic to our security, they are not the only risks. U.S. inferiority would present serious dangers if it existed in a number of areas of advanced technology already identified. For example:

Highly automated, smart, robot-like weaponry to carry out many air, ground, and sea warfare assignments, replacing military personnel and conventional fire-power. Here, our being ahead would compensate for the advantage the Soviet Union enjoys in numbers of soldiers.

Electronic warfare, in which we code our signals, detect the enemy's presence, jam enemy communications, spoof their missiles, and deny them use of their weapons by impairing their control radars and guidance signals, while blocking or misleading their efforts to observe and interfere with our electronic operations.

Space warfare techniques to offset a possible attempt by an enemy to deny us the use of space or to destroy our space systems.

High-energy focused lasers or ionized particle beams to replace conventional arms for destruction of land, air, ocean, or space targets.

Hardening of our domestic infrastructure to reduce the vulnerability to sabotage of our communications, transportation, electric power generation and distribution, water supply, and other vital systems.

Means to neutralize potential biological warfare weapons.

Underwater detection and tracking techniques which could radically reduce the present relative invulnerability of ocean-based nuclear weapons systems.

In summary, in allotting our resources to ensure the future security of the nation, the government has to preserve technological strength in the long term by maintaining adequate research capability and activity. At the moment, we do not have a proper emphasis here.

Schools and Security

If broad scientific research were recognized as mandatory for long-term national security, this would be readily evident in the high priority the nation assigned to the very foundation of scientific strength, namely, science and mathematics education in the nation's primary and secondary schools. The actual priority is low.

The typical American grade school student spends only one hour on science and four hours on arithmetic every week. Only a third of our high school students take three years of mathematics; most of the remaining two-thirds receive only one year, which means no mathematics beyond algebra. Only about 100,000 U.S. high school students study calculus in high school, usually for only part of a year, while 5 million high school students in the Soviet Union take a full two years. More than half of all U.S. high school students have not had even one full year of science, and no more than 10 to 15 percent receive instruction in physics.

Half of those teaching math and science in our nation's high schools are officially unqualified, are known not to possess the minimum required education for the task, and are engaged in teaching these subjects with temporary emergency certificates. This is not surprising since the training of both mathematics and science teachers declined by about 70 percent during the 1970s. Only one competent science teacher is available and licensed to teach science for every two high schools. In the last decade, shortages of equipment and supplies have forced high schools to cut by more than half the exposure of students to any kind of science laboratory experience.

Naturally, our high school students seeking to enter college are getting much poorer scores in the math and science sections of the college entrance examinations, the so-called Scholastic Aptitude Tests (SATs), than the students of a decade ago. The fraction of students who scored less than 300 points out of a maximum of 800 rose 40 percent between the middle 1960s and 1980, and the average score dropped by 90 points. Exceptional high schools are to be found in scattered parts of the country, but they are not sufficient for the nation to hold its own. Around 40,000 freshmen engineering students need remedial mathematics before they can even begin calculus, which they should have started in high school.

American grade school and high school inadequacies in science and mathematics are leading us toward a national security crisis. Today our schools are turning out science and math illiterates, and recent studies show that 20 percent of U.S. high school graduates are generally illiterate. This is a situation in which weakness builds on itself. High school graduates who lack a feel for science and technology and for the impact of these intellectual disciplines on society become voters who later, in their ignorance, will countenance even worse mistakes in issues that involve science and technology than their parents are now allowing our government to make.

The nature and importance of this matter is not one readily understood by the average voter and the local school board. We should not expect it to be. It is the federal government that must be the prime stimulator of special math-science activities, because the need can only be appreciated fully at the federal level—in just the same way that the need for ICBMs cannot be evaluated and rated against community needs by state, county, or city leaders. The school board of a typical city may be superior to Washington in determining and directing most aspects of the education suitable for the young people of its area. But relating education, from grade school through graduate school, to national security is another matter entirely. Here the federal government must not avoid taking on the responsibility for leadership in assigning the proper resources. On this matter we have not acted so as to ensure long-term national security.

President Reagan has been quoted at education conventions as agreeing that the education of American school children in science and mathematics has reached such a deplorable state that it threatens to reduce the nation's future military and economic strength. Yet the federal funding of such education has been lowered. In response to the Sputnik surprise in the late 1950s, the Eisenhower administration got Congress to pass the National Defense Education Act, and funds for education were provided to the National Science Foundation. The present administration has drastically cut the already dwindling annual NSF education funds from $80 million to $15 million, all of which will go to colleges.

As a funding alternative, private industry has been urged by the government to help local communities make up for decreased federal sponsorship of science and mathematics education in elementary and secondary schools. However, the industry leaders, people whose specialty is running companies and not nations, have promptly displayed essentially no interest in increasing their budgeted philanthropy. And why indeed should they assume that role for the nation? The costs would simply have to be passed on to industry's customers, who in the end are the taxpayers. Why drag in the corporations as go-betweens to collect the public's funds and distribute them as they see fit to schools to fill a fundamental need of the public? As the national security leaders in the federal government are the first to know, the benefit of the expenditures is to the nation at large. That is why we collect tax monies for defense purposes at the federal level and direct expenditures from that level.

The administration has established that top priority, indeed, crash efforts, must go to solving our economic problems and increasing our defense strength quickly. But long-term economic health and long-term national security deserve equal priority because our economic problems will not be solved and our security needs will not disappear in a few short years. In cutting budgets to achieve short-term economic improvement, we should not inadvertently arrange for inferior and limited education and a drop in the fraction of the GNP devoted to basic research. Both shortcomings are laying the groundwork for lower economic strength and impaired national security in the not-so-distant future.

Let us turn to college education. Starting some five years ago, a distinguished group of science and engineering leaders finally arranged for a substantial increase in competitive National Science Foundation fellowships in science and engineering, scheduled to take effect in 1981. Unfortunately, the present administration's budget not only removed the increase, it eliminated all new starts. Congress then restored the original low figures. It is hard to find a better indicator of our nation's neglect of its future security.

When university backing in engineering and physical science becomes highly inadequate, many of the best professors desert the universities for industry. Even if they remain, they devote less time to scholarly research and close work with Ph.D. candidates and more to hard selling to bring in scarce research grants or to meeting the added bureaucratic administrative requirements the government has created for research contracts. When the support is lowered by only a few percentage points, then perhaps the quality of graduate education and research decreases by only a similar small degree. But if the available funds drop by, say, a third, then, even if the graduate program of education and research does not collapse completely (which could well happen), the quality will descend to mediocrity. The situation is close to such a catastrophe in many university engineering and science schools.

We can launch larger defense programs at the expense of our commercial economic development and our competitiveness in the world's markets and still survive. However, if we lack enough people with expert know-how to implement soundly the contracted-for military projects, those projects will fall far short of meeting their objectives. Already our recently augmented defense programs, depending for success as they do on employing skillful scientists and engineers and well-trained technicians in the armed forces, are exposing that we lack adequate seasoned professionals to spend the funds efficiently.[1] Program time slippages, cost overruns, and performance disappointments, already common and expected, will steadily get worse year after year unless we recognize the growing shortage of expert personnel as a restraint building to an emergency.

The Department of Defense is now making a competent effort to enlarge its own sponsorship of research in the universities and to encourage America's large military technology industry to increase its university support. However, when the DOD presses industry leadership to use its own funds to improve the strength of the universities from which the industry will obtain its future staff, the response of corporate management is usually disappointing. A typical attitude of industry is that it is happy to provide more university aid if the cost will be included by the DOD as allowed reimbursable expense under cost-plus contracts. In other words, the industry says it will spend its own funds to rescue the universities if the DOD will restore that money right away.

---

[1] Military contractors, anxious to enhance their backlogs, will deny that this statement is true concerning their own ability to perform and take on even more projects, but they will profess it to be a quite accurate description of their competitors' situations.

Of course, some industry leaders have made positive suggestions for cooperation with DOD in enhancing American universities. For instance, they have proposed that industry should bring universities in to share efforts on their military R&D contracts, assigning to the faculty (perhaps assisted by graduate students) some of the more fundamental and unclassified portions of the tasks to be performed. If the DOD, when judging competitions for R&D contract awards, were to give points to contractors who do this, it would help greatly to motivate industry to provide subcontracting that would fund suitable university research and provide financial aid to graduate students.

Similarly, it has been proposed that industry sponsor symposia jointly with universities to survey the fundamentals of frontier research fields, and that the expenses be regarded as an investment in the future by industry. If all technological industry were to do such things, their competitive prices would all rise together and DOD would end up paying for the effort even as the industry appeared to be taking the initiative and using its own funds. Yet this is as it should be, because the DOD has the responsibility for national security, and that includes laying a foundation at the universities. It is right for taxpayers to pay the cost.

A program to improve the engineering schools of the nation and encourage a flow of doctorates in science and engineering is as critical to us in the long run as a sufficient production rate of airplanes and missiles, because eventually a lack of the former will ensure the failure of the latter. The entire cost of a major increase in the national budget for education and scientific research (secondary schools plus universities) could be financed by subtracting a small increment from our enormous budget for weapons systems. This would add only slightly to our short-term security risks, but it would build a solid foundation for the years to come. Not to do so guarantees greatly increased security perils in the future. We shall return to this topic later to propose additional actions. But first it will be helpful to discuss further the use of advanced technology in warfare.

Technological Advance and Military Strategy

The status of our science and technology will always dictate the weapons capabilities the military can hope to possess. However—as with the conundrum of the chicken and the egg—security strategy should include stimulating basic research and technological advance to take us beyond the existing weapons art. The NATO situation exemplifies this. We badly need a new nonnuclear NATO strategy, and we need science and technology to match that new strategy.

The NATO pact, which unites the efforts of western European and North American countries with a common interest in the defense of western Europe, has led to a contingent of some 350,000 Americans in Europe. The presence of American troops has been a symbol and guarantee of American commitment. The NATO military forces possess nuclear arms, but the steady buildup of conventional Soviet forces in Europe implies that the invading armies would have more soldiers, tanks, airplanes, and overall conventional firepower than the total of the NATO forces, absent the nuclear weapons, opposing them.[2]

Still, even relying only on conventional military defense, NATO forces might inflict substantial casualties against incoming armies, and this possibility has been considered a significant deterrent to a Soviet takeover. However, the heart of the deterrence has been NATO's avowed willingness to employ nuclear weapons, if necessary, to defeat the Russians.

Unfortunately, the foregoing is a description of an out-of-date strategy. A new understanding developing in western Europe and America is that nuclear weapons will not be used first by NATO forces, because that would bring quick retaliatory nuclear blows by the Soviet Union, and Europe then would become a nuclear battleground, a condition western Europe's citizens are expected to make clear they are unwilling to allow. Thus a different approach is needed for the defense of western Europe, one that assumes nuclear weapons will not be employed and that the Soviet Union (their own first-strike nuclear option equally deterred by inevitable retaliation from us) will use only conventional forces should they decide to occupy western Europe.

What should be the essence of NATO's new approach?

A sensible answer must first take account of the relative inherent strengths of the opposing sides. The nations of western Europe, even with no American participation but with the will to resist the Soviet Union, seem capable of deterring a nonnuclear attack and of winning if deterrence fails. Western Europe has around twice the GNP of the U.S.S.R. and is capable of more equipment production. It has a population base that makes it statistically possible to field an equal number of soldiers (especially if one assumes a

---

[2] For some years the U.S.S.R. has been producing twice as many aircraft and tanks as the United States. A DOD report to the Senate in 1982 stated that the Warsaw Pact armies would have roughly a two-to-one advantage in overall firepower across the entire battlefront.

reasonable assignment of Soviet forces to guard the Chinese border). Western Europe exceeds the U.S.S.R. in overall scientific, engineering, and industrial strength. The NATO nations of western Europe have about 250 million inhabitants with a credible claim to considerable mutuality of interests. If united, they should be a match for the 80 million or so Russians together with about 170 million non-Russians within the U.S.S.R.[3]

The Soviet Union probably has more natural resources than western Europe, even with Canadian and U.S. resources included, but the infrastructure to make most of those resources available is underdeveloped. In the next few decades western Europe alone, if it wished, could arrange to place more resources from the world as a whole behind its military effort than could the U.S.S.R.

NATO consists of independent nations that find it difficult to integrate their forces into a highly coordinated defense body. However, other nations of the Warsaw Pact cannot be regarded by the U.S.S.R. as happily devoted, loyal teammates. Of the fifty-eight Warsaw Pact divisions, fifteen are Polish, who in an invasion of western Europe would probably display an enthusiasm to die for the Russians that would start low and diminish rapidly. Ten divisions are Czech, and experts often list them as the least dependable of all Warsaw Pact divisions. Thus a large fraction of the total forces that presumably would mount a land attack against NATO nations might not be assessed by Soviet military planners as possessing guaranteed effectiveness.

Considering these facts in isolation, one might conclude that western Europe, if it chose, could present at the least the probability of a stand-off to the Soviet Union's presumed conventional attack. This should constitute a real deterrent because the Soviets would not find it attractive to start a war that they could not win quickly. A stalemate turning into a war decided by production rate would be tantamount to their losing. However, the immediate situation is different. With a plan to use nuclear weapons as the true counter to an invasion, 350,000 American troops in Europe are too many. On the other hand, if nuclear weapons are ruled out and we assume the need to stop the larger Soviet armies with conventional forces, then these American units plus the western Europe armies add up to too few.

Are western European nations interested in greatly expanding their conventional forces, in arming adequately to resist a Soviet military takeover? If not, even though they have the basic capability to do so, this means they

---

[3] In 1980 the population of all the Warsaw Pact countries was about 380 million and that of all of western Europe, including the nonmembers of NATO, was around 430 million.

lack real commitment. Without adequate resolve, their creating a conventional military force that is fully as large as the Soviet forces, with maintenance
costs which would be severely penalizing economically, is an unrealistic
objective. The continual need to accommodate Soviet pressures would wear
down and eventually overcome a weak determination to maintain so
burdensome a defense budget. The Soviet Union, as a consequence, would
ultimately prevail in Europe—if not militarily, then through the growing
influence of sympathetic political leadership in western European countries.

However, a more innovative approach is possible. If future strategy is to
be based on assuming no nuclear bombs will be employed, it is reasonable to
ask whether a scientific and technological effort to develop appropriate new
weapons systems is not a far more suitable role for the United States to play in
western Europe's defense than placing larger American armies there. How
can advanced science and technology best fit into the act? In what areas of
pertinent science and technology are the United States and western Europe
ahead of the Soviet Union? What kinds of new tools can we develop and
produce to give our side a real edge in a nonnuclear war?

The answers lie in the full utilization for military purposes of advancing
electronic technology. In computers and communications systems, semiconductors, microminiaturized solid-state electronics, microprocessor chips,
etc., we are far ahead of the U.S.S.R., and there is every reason to believe we
will outdistance them in the future, especially with Japan properly included
on our side. With all-out employment of this new technology, highly focused
military action is made possible, a result of superior command, control,
intelligence, and reconnaisance. This superiority would be based in part on
our possessing more accurate and complete military information. At all times
during military engagements we would be able to use our better information
technology to know what is going on in detail and make that knowledge
useful to direct our military efforts in optimum ways.

We would disturb Russian communications while preventing them
from hampering ours. Coding, jamming, intercepting signals, processing
incoming data in real time, making analyses, and presenting alternatives—all
typical offerings of added electronic brainpower—would make our forces
smarter and better led, positioned, and used, while we would befuddle theirs.
Our forces would become much more powerful ones.

Our higher technical status in relation to that of the Soviet Union in the
electronics of information handling enables us to design intelligent robotic
weaponry of high effectiveness. A comparison of our cruise missiles with
Russian versions is highly illustrative. Theirs are crude, heavy, and inaccurate
relative to ours, even though we have hardly begun to inject our latest

microelectronics technology into the design of these missiles. We are many years ahead in cruise missiles and should remain so.

Soviet tanks now outnumber NATO's by a factor of more than four. However, defenders who do not intend to become attackers have less need for tanks. Tanks to resist tanks is a concept of warfare rapidly becoming obsolete. Smart, lightweight, mass-produced, accurate guided missiles that hunt and close in on fixed or moving targets are the cause of the obsolescence. Hand-held, aircraft- or helicopter-launched, or artillery-directed missiles utilizing laser, microwave, or infrared beams can act across the path of a tank assault as it develops. Advanced homing electronics can enable an artillery piece to launch a smart bomb that can hit a moving tank directly from 10 miles away. Huge volumes of directed ground-to-air, ground-to-ground, and air-to-ground projectiles can be launched from a distance in safety, because technological advances permit us to "fire and forget" these target-seeking warheads.[4] These new hardware advances will lead to a defense that is inherently stronger than offense in land-takeover battles.

Although the Russians have spent billions of dollars on air defense, our electronic countermeasures and jamming could disrupt the operations of their systems, and the decreased vulnerability of our missiles and planes to detection could make Soviet countermeasures less effective. The use of the air could be denied to them by the vast guided missile population we could create, while the ground terrain could be made further impenetrable to them by brainy electronic mines, set off when the enemy sought to enter an area but rendered harmless to our forces by broadcast of coded instruction signals.

Required of us in implementing these potentials of our technology would be the engineering and high-volume production of precision electronic and mechanical equipment. This is exactly where America, western Europe, and Japan are very strong and where the Soviet Union is very weak. The Soviet Union's relative technological backwardness in this kind of mass apparatus production would suggest both a qualitative and quantitative imbalance in our favor, just as if they were to attempt an invasion with fewer and less well-equipped forces than ours.

What we are describing here is not to be confused with the idea of

---

[4] One antitank missile system, called the Assault Breaker, has been described in public literature. It is designed so that if our radar discloses enemy armor massing, a carrier missile is launched which when over the area spews out dozens of small missiles, each capable of sensing and homing on a tank. One such carrier missile could take out an entire company of tanks over 25 miles away.

merely adding complexity to conventional weapons systems to outdo the Russians, who presumably will have only simpler weapons. It has happened in the past that in determinedly seeking improved performance, U.S. military services have loaded too many added features onto basic weapons. This has made them so complicated and unreliable under battle conditions as to defeat their purposes. On the contrary, like American computers that are getting smaller, cheaper, smarter, and more reliable each year, the electronically controlled weapons we refer to will be more economical and dependable in mass production.[5] This is because the sophisticated functions of which they are capable are inherent in the miniaturized semiconductor circuitry on chips, the product of major breakthroughs in the techniques of electronic fabrication.

Military experts on the problems of an invasion of western Europe from eastern Europe point out that there are only four usable routes, none of which is easily traversed because of the disadvantageous physical aspects of the terrain. These paths include canals, bogs, concentrated urban sprawl, and even some narrow, winding mountain passages. Basically, the maneuvering requirements for an attacking force are more difficult than those for a defensive force. The Warsaw Pact nations would have to concentrate their forces to seek greater power for the spearheads they will be striving to push forward. They would find it hard to manipulate their numerically populous divisions to attain a predominantly superior force at the point of battle because the routes do not lend themselves to adequate maneuverability. With our strength in information technology making our reconnaissance superior, we would learn what they are doing quickly and could then focus our directable, technological power on the high-density attack centers.

Also, for a successful invasion the Warsaw Pact armies would need excellent command and control to enable their military leadership to seize opportunities of the moment, shifting forces and tactics as necessary. If their command and control were interfered with by superior NATO electronic warfare capability, then the Russians would find their attack bogged down and bottled up, and their higher numerical forces would make counterattacks by automatic, smart weapons all the more effective.

---

[5] A present microcomputer, compared with an early electronic digital computer, costs less to produce by 1,000 times, is 100 times faster, uses one-thousandth of the space and power, and is 10,000 times more reliable (that is, will run 10,000 times longer without a failure).

The Political Weaknesses of a Smart-Weapons Strategy

What we have just described, the application of advanced technology to back up a nonnuclear strategy for the defense of western Europe, is a natural for NATO, technically and economically. But in other respects it is anything but natural. It presents some formidable political difficulties. One is in organizing for cooperation among all the member nations of NATO. The advanced technological devices we need to develop and put in high production for NATO's military needs have great similarity to the latest advances in commercial computers, civilian communications equipment, factory robots, industrial controls, and evolving control apparatus for the newest airplanes and automobiles. The nations of NATO are the toughest of competitors with each other, the United States, and Japan for the commercial exploitation of this very technology in the world's civilian markets. The need for the military technology is clear, but to obtain the necessary teaming among these competitive nations may be an insurmountable challenge.

What would we have to do to attain the necessary close cooperation? First, employing advanced technology to provide NATO with important defense advantages over the Soviet Union will fail if such action is not accompanied by broadly harmonized U.S. and European interests. All NATO nations would need to agree, for instance, on the same high priority for military applications as compared with commercial product development. If, in contrast, the Europeans are intent on building up their economy in part by increased export to the Soviet Union, including shipments of civilian information technology equipment, they would be pursuing a foreign policy strategy not only different from ours, but one that would undermine our basic technological advantage over the Soviet Union.

Next, we would have to overcome a significant organizational flaw that NATO has suffered from its beginning. The member nations have been willing to give up so little autonomy that no really authoritative central command over the strategy for the combined military forces of NATO exists. (We must also recall that France, although cooperative, elected years ago not to be a formal member of NATO.) This weakness has been somewhat camouflaged in the past by the plan to rely on NATO's employment of nuclear weapons as a deterrent. The United States has control of the nuclear weapons that it supplies to NATO, and the overall U.S. nuclear capability is very much stronger and more broadly based than that of the British and French forces (while Germany and the other nations possess no nuclear capability). This nuclear arms superiority has the put the United States in a nearly commanding position regarding deployment of nuclear weapons in Europe. However, if we are going to shift now to a nonnuclear defense strategy, then the lack of a tight, single command of all western European forces becomes much more conspicuous. This is especially true with the new electronic weaponry approach.

To engage in electronic warfare and employ robot-guided missiles or smart weapons absolutely requires the development of a single integrated system of hardware, software, fighting equipment, personnel, intelligence, reconnaissance, communication—with total control of them all at all times. Otherwise, we might not only fail to attain the needed focus of our defense, we might be in great danger of jamming and even attacking our own forces. The system must bring together, process, and disseminate the data as to what the Soviet forces are doing and direct the response in a near optimum manner. Success in this kind of warfare is not compatible with a fragmented organizational command structure.

Finally, it is exceedingly difficult to solve the problem of attaining close cooperation and tight central control because the west European nations appear to hold views substantially different from ours as to the seriousness of a potential Soviet invasion. They often act as though they thought such an action by the U.S.S.R. extremely improbable. They sometimes look upon their armed forces not as a critically necessary guarantee of their security but rather as a source of employment and industrial activity, to be taken advantage of and dealt with as they would any other economic opportunity. The effort required by their technological industry to support their military forces is seen as a sort of adjunct or stimulant to their normal civilian industrial activity and a source of government funding for technological advance and employment.

They are quite willing to gamble that it is advantageous for them to work out stronger trade ties with the Soviet Union and to take financial risks in backing such trade. They see this as a way to increase the backlog for their industries and improve the health of their internal economies, even as their output helps the Soviet Union to develop its natural resources. From their perspective, western Europe and the U.S.S.R. are merely seeking long-term peaceful relations which will cause them both to prosper. As they compare overall risks, it does not seem sensible to a good portion of the leaders and citizens of western Europe that they give up autonomy and economic growth to make possible a highly coordinated military defense against a Soviet invasion which they regard as improbable.

No matter how science and technology are to be used in providing security to NATO members, this basic difference in assessing the Soviet threat to western Europe has to be faced. Granted this limitation, and putting it aside for the moment, we must note that we are not today pursuing as energetically as possible the buildup of the type of NATO defense earlier described, i.e., defense based on robotic, smart, electronically guided weaponry. Not only are we slow in mounting the full-scale development of the electronic warfare age, we are continuing to be wasteful in a continued overemphasis on the nuclear deterrent to a Soviet attack.

Thus Britain is planning to spend several billions of dollars on an independent force of Trident submarines carrying long-range missiles with nuclear warheads. Now it is understandable that Britain's military leaders would seek for their inventory the most technologically advanced strategic weapons. Still, any circumstance in which Britain would use those Trident submarines independently is impossible to imagine. As still another contribution to the total force that our side can display to deter the Soviet Union from starting a nuclear war, those billions invested by Britain are somewhat helpful. But it would be far more sensible if they would assign their limited funds to smart weapons, nonnuclear in nature, stationed in western Europe, thus making use of the latest technology to deter a Soviet nonnuclear invasion.

We even may have to face the fact that there is no realistic way to maintain the present NATO alliance. Unless we can fit advanced technology into the military roles and missions of the commercially competitive nations that make up NATO, we cannot put up a joint defense that is credible for deterring the Soviet Union in the nuclear-free era ahead. If the United States withdraws from NATO, this does not mean that we cannot join with western Europe in a different, perhaps more suitable alliance that will add greatly to western Europe's ability to deter Soviet military takeovers. For example, in leaving the western European nations to arrange responsibilities, expenditures, and strategy as they see fit for defense of their countries, we can deed to them the U.S. nuclear weapons now there.

With us absent, the western Europeans could decide by themselves what size and kind of military effort they wish to pay for, making their own estimates of the risks and negatives of losing a war versus the economic gains of holding down defense expenditures. They could decide on the role of nuclear weapons without concern as to how the United States and the Soviet Union might connect strategic nuclear exchanges with local European hostilities. If they want smart weaponry and other advanced electronic gear from the United States, we should consider supplying it to them. We even might elect to give them some of this equipment free of charge if we felt that their strategy for the defense of western Europe made sense and was based on adequate determination and conviction, and if they agreed to safeguard any classified technology they obtained from us. We could then find it worth the cost to the United States to see western Europe remain independent of the Soviet Union. We should not accept any proposals that we donate equipment on the grounds that without such aid western Europe cannot afford to defend itself. It obviously can. If it does not wish to pay the price to do so, then contributions in technology from us will not ensure victory.

The Shortage of Long-Range Strategy Planning

The NATO example shows conspicuously how security for the United States involves more than superiority in technology or military force and that for success, numerous difficult economic and political issues, domestic and international, must be integrated in a full systems approach. The DOD, it must be recognized, does engage in considerable long-term planning. However, this mainly is in response to the practical situation that weapons systems typically use up five to ten years from the start of R&D to operational capability. It is less the result of a deep commitment to establish broad, long-term strategy. Overall U.S. planning of a kind that transcends the detailed plans and efforts of the individual military services is much too weak. For instance, the highest-level position in the DOD assigned full-time to pondering the important strategy problem of cooperation with our partner nations in NATO is that of the deputy assistant secretary for NATO affairs. As of this writing the job is unfilled and has been vacant for two years.

Failure on the part of western Europe and the United States to engage in adequate, broad, long-range strategy deliberations constitutes a fundamental limit to our exploitation of technological advantage to provide for our security. This inadequacy stems largely from the fact that the democracies of the West are plagued by government leaderships that have short lives. Government policies that are visible are based upon crises of the moment in an effort to be responsive to the short-term thinking of the voters. The governments generally have no real long-term designs as to security, except for the desire to be free from takeover by the Soviet Union.

In contrast, since the beginning of Communist control of the Soviet Union early in the century, their heads of government have been few and their periods in office have been long. While France has had to adjust to ten times as many changes of governments in the same period, Soviet foreign affairs have been under the same individual for decades, and the same generals and admirals are seen for long periods at the head of their military operations. Soviet strategic and military doctrines, while perhaps steadily refined to adjust to technological advance, have a solid core of long-term stability. Moreover, the Soviet Union works hard at mastering the art of combining the nontechnological forces available to them with the technological ones to further their aim of getting their way in the world. They do not have the problem we are now discussing of ensuring a proper match or balance between what they do on the military or technological front and what they seek on the political and propaganda front.

The chief U.S. security policymakers, such as the president and the secretaries of State and Defense, seem to average four years in their posts and leave them as they begin to learn in detail the real parameters of their jobs. Only a few of the top hundred civilians whose positions require them to lead in shaping major security decisions or recommendations to Congress have long backgrounds in defense matters. (Secretaries, under secretaries, assistant secretaries, deputy assistant secretaries, etc., of the DOD and the armed services and the staff of the National Security Council are typically replaced wholesale when an administration changes.) Few holders of these slots have had the opportunity in their previous careers to study security strategy alternatives in depth. A good part of the first term of each administration is always over before its leadership has been able to progress in security strategy from positions taken during the election campaign, and such positions are bound to have been influenced by perceptions of what will bring political success.

U.S. Defense Department leaders have been quoted in the past as saying that the United States does not have a coherent set of plans for conventional (nonnuclear) warfare at sea, on land, and in the air, nor have adequate plans been made to ensure the mobility of our forces. General David C. Jones, upon leaving his position as chairman of the Joint Chiefs of Staff, was particularly outspoken on these points, saying that unless the defense establishment itself is reorganized, the American people will not get the defense capability to which they are entitled and which is needed. In particular, he referred to the need for a new form of Joint Chiefs of Staff, one that could rise above interservice rivalries and bureaucratic resistance to change. At present, of the five members of the Joint Chiefs, each one except the chairman wears another hat, namely, as the chief of his organization. Each of these chiefs, although a member of the Joint Chiefs, has a heavy responsibility for meeting the goals and allotted tasks of his service unit. He is briefed by the staff of his separate service on how that entity sees requirements and strategies. The Joint Chiefs of Staff has a staff of its own, but all such staff members retain their memberships in and loyalties to their respective services, to which they look for promotions and advancement in their careers.

Almost every retiring chairman of the Joint Chiefs has made comments similar to General Jones's, and the same thoughts can be gleaned from the statements of those who have held high White House positions related to security. We are told (after they leave office) that when in those posts, their efforts were completely taken up with immediate crisis matters. Under such conditions it is understandable that inadequacies exist in longterm strategic thinking and in considering interactions among the technological and nontechnological factors that could provide solid national security in the future. Former staff members have stated that the National Security Council itself has been more a vehicle for displaying decisions than an instrument for arriving at them. It is not, as outsiders often assume, a forum for the president and top officials to discuss major policy options influencing security and to figure out how security decisions affect and are affected by domestic spending, trade, energy, and other policies.

Similarly, the staffs of those congressional committees concerned with defense matters, even though they have become very large, do not devote much time to long-range security strategy. Instead, they are almost exclusively involved in detailed oversight of DOD's existing programs and proposals for the immediate future. The congressional contribution thus is sporadic, spotty, sometimes arbitrarily so, and focused on assignment of funds to individual defense projects.

No wonder our policies with regard to national security, when they are articulated, are observed to oscillate between extremes. They range from crash, emotional, overspending binges, in part on the wrong procurements, to periods of substantial neglect, at least as judged a few years later by the next administration.

Long-range security strategy for the nation cannot be expected to come from the private sector and surely not from the military technology industry. Of course, the industry employs many individuals capable of long-range planning and strategy formation in the national interest, but their jobs require instead that they push for a maximum of spending in their fields of endeavor, thereby ensuring a steady flow of contracts that will give their companies a good return on investment and growth of revenues. Such is not necessarily a growth matched well to the defense needs of the nation which, overall, is not industry's responsibility.

Rather than emanating from broad, long-range strategy, much of our defense policy arises almost accidentally from short-term budget considerations. For the year ahead, the overall budgeting process leads to a rough total for all government spending that is politically acceptable, then a subtraction from that total of essentially fixed items about which there is no real choice. A tentative decision can then be reached on the fraction of the whole to be assigned to defense spending. Almost everyone involved in the process has to concentrate particularly on the next year or two and on the allocations to detailed areas.

Policy questions are examined through the activities of the National Security Council staff, the State Department, the Defense Department, the Arms Control Agency, the Joint Chiefs of Staff, and the CIA—that is, when these groups are not fully occupied with near-term emergencies—but these deliberations usually are quite distinct from the more dominant budgetary influences. The budget, which is formed periodically, with minimum tie-in to whatever formal long-term strategy deliberations exist, comes close to controlling strategy by default, that is, by the accumulated effect of year-to-year, short-term funding decisions.

The Japanese Security Contribution

The NATO problem is but one illustration of the difficult political-economic complications that arise when attempts are made to integrate security with foreign policy and arrive at a long-term national security strategy. Deciding on how to get Japan to be more active in defense matters is another good example. The United States has never settled questions such as these: Should our efforts be restricted to persuading Japan to buy American arms in order to build a larger Japanese military presence in the Asian theater? Should the Japanese be pressed to supply NATO with a substantial fraction of its needed, mass-produced, electronically controlled, defense missiles? Do we prefer that the Japanese independently enter the worldwide arms business?

Japanese companies already make certain American-designed weapon systems for their modest local defense program and are said to be doing so with high quality.[6] However, Japan has a firm policy against producing

---

[6] It is said that radar equipment for the F-4 Phantom, when built by Japan, is rated as running for thirty-five hours between maintenance stops, while the same equipment made in the United States will work for only ten hours before a failure. Likewise, some Japanese versions of U.S. defense missiles have been judged superior in accuracy, the result of higher standards of quality control in production and maintenance.

military hardware for export, and it has stayed out of international arms trade. Do we really want Japan to get into that market? If we look at the various alternatives for Japanese participation in maintaining the security of the non-Communist world, quite different possibilities suggest themselves. All have substantial potential effect on the Japanese and the world economies and on the contest for superiority in civilian international technology as well as on security. Specifically, Japan's entering more deeply into weapons production can affect the economic health of the arms industries of western Europe and the United States, even as such an effort by Japan in the military field would make them a weaker competitor worldwide in purely civilian technological products, since their technological resources do have a limit.

Most of the technological systems that Japan would need if they were to carry a larger share of the defense mission for the non-Communist world is already being produced in the United States. Our large production volume in airplanes, missiles, radar, communications gear, and computers suitable for such military tasks lowers our production costs. Even though the Japanese are extremely efficient in manufacturing technological apparatus, it would cost them less if they purchased their needs from us because their low-volume costs would be higher and the start-up expenses would heighten their costs even more. If Japan were to transfer funds to the United States to purchase such military hardware, they probably could make up those funds by concentrating their own technological resources on the civilian commercial field. An alternative would be for us to assign them the production of some mutually needed items and for us to purchase from them to cover our requirements. The idea would be to obtain standardization and the best weapons systems with the least expenditure by each nation.

It happens that Japanese-U.S. trade relations are now being increasingly impaired by the imbalance of trade that exists between the two countries. The Japanese cannot continue to enjoy large favorable trade balances with the United States without eventually triggering U.S. protective restrictions which will handicap both nations. The military needs for equipment of both nations should be introduced into this equation with the aim of finding the most satisfactory overall package for good commercial trade relations and maximum security with the least investment and annual expenditures.

Instead of analyzing all these matters—developing long-term preferred options in which economic and political as well as military security aspects have been integrated—and then proposing these options seriously, we merely apply ad-libbed criticism and pressure on Japan from time to time. We utter hit-or-miss slogan-type complaints about their not carrying their share of the load we envisage as necessary to deter potential Soviet aggression.

Space Warfare

As a final example of important security issues which involve science and technology, but where the political-economic factors are even more challenging than the technical ones, consider space warfare. In an earlier chapter, we called attention to the fact that both the United States and the Soviet Union consider it essential that they utilize space for military purposes. We listed communications, command, and control of strategic weapons, intelligence, reconnaissance, and early warning as requirements for the space race. We cited the indispensable need for inspection satellites to make practical and credible any agreements that might be set up for reduction of nuclear arms. We also noted the fact that since equipment in space offers advantages to the military, a potential enemy will consider developing the capability to destroy its rival's space systems.

Space warfare, then, is not to be ruled out. It is technologically feasible to destroy whatever equipment is placed in space. It is also technologically feasible to observe that an enemy is preparing to attack one's space components and to take defensive action against the attacker or to preempt such moves by attacking the enemy's space systems first.

Furthermore, it is to be noted that ICBMs, and even most sea-launched ballistic missiles, travel on trajectories which pass through near-in space. As far as technology alone is concerned, space systems can detect the launching of such missiles (and can also track airplanes, particularly those at high altitudes), and defensive attacks against the tracked apparatus can be launched from space-based systems. Thus, in thinking about future space warfare, we must consider the possibility of actions to interfere with a wide range of military activities that utilize equipment in space, while recognizing that warlike acts which start in space may not necessarily remain limited to space.

All of the technological requirements involved can be met roughly equally by the United States and the Soviet Union, since both now are strong and versatile in military space technology and have considerable background for the development of weapons suitable for the kind of space warfare we have outlined.

What about the economic aspects of space warfare as distinct from mere technological feasibility? The various systems range from some economically reasonable to others fantastically expensive. Hardly any other dimension of warfare conjures up more quickly the aspects of an expensive race with no end. This is because the moment one describes a system to destroy an enemy's space-based equipment, it is equally possible to propose a system to counter it. The first system then can be readily imagined as extended, refined, and developed further to make it more powerful and to allow it to protect itself against attack. Each successive system increment of offense and defense grows higher in cost and complexity. It is like two expert marksmen shooting at each other: the one that combines greater range and accuracy will kill the other. But then the second has only to acquire a gun with somewhat more accuracy and range and surround himself with armor and he will temporarily become the winner—that is, until his opponent adds more firepower and accuracy, etc., etc. In space warfare, the Soviet Union and the United States could easily find themselves in such a competition, spending more and more on research and development and procurement, seeking an edge in performance that would deny to the other the use of space.

We definitely do not need another economically wasteful contest. On the other hand, we cannot allow the Soviet Union to control space. The Soviet Union, for precisely similar reasons, while not relishing the idea of engaging in this competition, cannot countenance the possibility of U.S. domination of space. It is probable that if were to greatly reduce our present nuclear weapons systems by verifiable agreements, then we also would decrease the probability of space warfare. Having in all seriousness arranged to diminish the likelihood of a nuclear war, we should be able to agree that no hostile equipment should be placed in space and that space should be reserved for the use of both powers for intelligence and warning, arms-control inspection, and civilian commercial purposes. If, on the other hand, we cannot come to agreements about nuclear arms reductions, then the space race will come to override, in every sense, the present nuclear weapons race.

The highest of priorities must go to seeking very early agreements to "disarm" space, to avoid a military space duel. We need, that is, to commence a second arms reduction effort, even as we are only starting the first effort, the nuclear-weapons disarmament talks whose success appears to be a mandatory preliminary step towards the demilitarization of space. If we are inclined to be optimistic, we can argue that the enormous penalties to both sides of our indulging in a space-warfare race, when understood, may actually accelerate the nuclear weapons negotiations. What needs to be negotiated may be complicated, but it is possible to state it: starting on the second problem (the space duel) before the first (nuclear weapons) is solved may cause the first problem to be solved earlier and may make the second problem's solution much easier as an immediate follow-up.

Technology Transfer and the Military

We want to give our allies every aid our technology can provide to advance common security purposes. Yet, since we are competitive in civilian applications, we must limit the disclosure of our technology to our allies if we are to forge ahead of them in the world contest for technological superiority. Thus we face a dilemma as we try to work closely with our allies. It would seem, by contrast, that keeping our technological know-how from our potential enemies should be more straightforward and, at the least, should present no dilemmas. Not so.

Proper study of the history and status of Soviet military technology leads to the firm conclusion that they have benefited greatly from procured information about U.S. military technology. They have used spies, seduced American citizens with access to classified information into violating the law for money, and exploited disloyal Americans possessing military data who ideologically would prefer Soviet world dominance. In these ways they have obtained information ahead of its normal declassification and release. They also have taken full advantage of our basically open society, studied U.S. publications thoroughly, and attended pertinent American public discussions, university lectures, symposia, and conventions of scientists and engineers. Finally, they have acquired U.S. commercial equipment with applicability to military weapons systems.

An often-cited example is a sale by a U.S. company to the Soviet Union several years ago of a $150-million factory to build petroleum drill bits. The key technology here involved the use of tungsten carbide, an extremely dense and very hard metal alloy. Tungsten carbide cutting edges permit oil wells to be drilled through hard rocks; the same alloy is used by the military to penetrate armor. It is said that the technology we exported on this program has enabled the Russians to improve their antitank weapons. Similarly, precision ball-bearing manufacturing machines sold to them a decade ago are believed to have enabled them to heighten the accuracy of their intercontinental ballistic missiles.

Why did we not give these technologies a security classification and thus prevent their export? The practice in the United States is for the government not to try to classify everything important to our security because that would be a practical impossibility in view of the large number of facilities, equipment, and items of information to be included. Judgments about what to classify are applied thousands of times daily by individuals from top to bottom of the whole hierarchy of those handling information and equipment.

The newest, most sensitive, urgent, and dangerous (in the hands of a potential enemy) items are protected the most carefully. As time passes, bans on specific items are removed, and the items are allowed to become public.

Rules already exist to prevent the shipment abroad, particularly to the Soviet Union and other Communist countries, of equipment and design information in specifically listed technological areas such as lasers, semiconductors, and computers (whether the items are classified or not). The Commerce Department has the authority to enforce the government's regulations to restrict exports in such high-technology fields, and it issues commercial licenses for some $5 billion a year in technology and commodity exports. Since there are over 300 highway, air, and ocean exit points from the United States, the Commerce Department has hardly been able to do more than sample cargo and cannot know how much leaves the country illegally. It has been estimated that some tens of millions of dollars of semiconductor products and data are purchased annually for illegal export from California's so-called Silicon Valley alone. As the 1980s opened, the Commerce Department had a backlog of several hundred incomplete investigations involving allegations of unlicensed exporting of technology.

The State Department issues export licenses for billions of dollars a year in arms and other military equipment, its area of authority. Again, only spotty and relatively modest physical inspections are made of exports to identify improper ones, since the State Department has neither the funds nor the expertise to examine all shipments of high-technology equipment that may contain illegal items marked and packed so as to camouflage their real nature. As an example, one small nation, Libya, which finances terrorist activities all around the world, continually engages in clandestine efforts to obtain U.S. weaponry. Complete checking of all shipments to Libya alone, directly or indirectly, would probably use up the efforts of the entire State Department investigatory staff assigned to this problem. Besides, regulations on exports do not even cover thefts that have a covert intelligence foundation.

In attempting to control the flow of sensitive technology to potential enemies, the United States is faced with conflicting desires. One is to sell as many goods as possible to other countries, and the other is to deny them access to our technology for reasons of both national security and commercial competition. For every single American or government bureaucrat or member of Congress who wishes to hold back an item from export because it may aid a potential enemy, there is another individual or company or government leader who will press to export that very item to increase foreign trade. There has never been a really strong effort to resolve this dilemma. Instead, we have constructed complex and cumbersome rules and organizations intended to control technology transfer while simultaneously trying to appease forces opposing control. The consequences are often that neither the assigned government administrators and inspectors nor the industry find they can deal effectively with the maze of confusing directions.

Several times a year meetings are held by government and industry representatives to try to agree on the procedures and responsibilities. The attendees mainly spend their time at such sessions calling attention to the myriad of details of regulations under which they operate in frustration. Even when requirements are reasonably well understood, everyone concerned complains of (1) being hopelessly overloaded with the volume of research required to unearth the basic data so that decisions can be made and (2) having a volume of traffic far greater than anyone can handle.

A good example is the attempt to control U.S. machinery used to produce semiconductor chips, the basic building blocks for the precision microcircuitry that has revolutionized electronics. Over a quarter of all chip-making equipment produced in the United States is exported to friendly nations, presumably none directly to the Soviet Union and other Warsaw Pact countries. However, our NATO allies do not regulate the resale of this equipment on the so-called used market, where some of it seems to show up very soon after its original export and which is accessible to purchasers from anywhere. It is estimated that in the last decade the Soviet Union has purchased hundreds of millions of dollars of U.S. equipment for semiconductor production through this route and by more clandestine means, such as the use of phoney front companies in the United States and western Europe.

As another example, consider our relations with the People's Republic of China. Some parts of the U.S. government are promoting relaxed restrictions on transfer of military technology to China to make that nation a semially as a counter to the U.S.S.R. Also, the leaders of our National Academy of Engineering have met with their counterparts in China, and they are arranging for Chinese engineers to take year-long leaves of absence to work in U.S. technological industry so as to be advanced in their knowledge and training. Other U.S. government units are encouraging the sale of commercial civilian equipment to China. This encouragement, however, carries with it the warning that such sales are forbidden if the equipment has "potential military applications." Thus, a U.S. home computer, which can readily be bought by foreigners in the United States and later transferred to China, may be embargoed for direct sale to that country because the computer uses semiconductor chips. These, after all, might be applicable also to military apparatus. (Or it may not be embargoed; it is not certain which.)

Meanwhile, Japan, presumably one of our allies, is not bound by these rules and can sell semiconductor chips freely to China, as do Hong Kong manufacturers, most of whom got their start with technology licensed from the United States years before. As these words are being written, U.S. computer companies are considering accepting an invitation to bid on a $200-million computer order from China, financed by the World Bank, but no one in industry and few in government feel certain which items on the purchase list may end up being banned for export because they might be regarded as of potential military aid to China.

As a final illustration of the confusion, the United States strangely decided recently to allow the export to Germany and France of novel laser technology used to separate isotopes. The application of this technology is to produce highly enriched uranium and weapons-grade plutonium at costs expected to be well below older methods. When developed, this technology could proliferate nuclear bombs if it fell into the wrong hands.

The U.S. government is now considering additional restrictions on outward flow of information in an effort to handicap the Soviet Union more.[7] The new proposals would constrain the degree to which valuable advanced scientific and technological concepts and experimental data and the like are exported or even discussed publicly, published, or distributed within U.S. boundaries. Specifically, it is an attempt to prevent Soviet citizens from attending meetings in the United States on such subjects.

However, severe practical difficulties abound in endeavoring to control the flow to the Soviet Union of all U.S. civilian equipment and information that conceivably might end up aiding their military effort. First off, classifying a larger fraction of all science and technology involves enormous additional expense. More important, as we have already indicated, is that restrictions broadening the classification process can be a great impediment to necessary communication within the United States among those who are engaged in advancing our technology. The more we break up information and assign it to closed-off cells that are not in free communication with others, the more difficult it is to generate and maintain expertise. It takes a large team of constantly conferring specialists in science and technology to move rapidly

---

[7] Bills have recently been introduced in Congress to create a new agency, the Office of Strategic Trade, that would take over the control of technology transfer to other nations. The proponents argue that the departments of both State and Commerce have conflicts as they seek to limit export of technology for security reasons while simultaneously promoting it for reasons of trade or diplomacy.

from research and development at the frontiers to the design of new weapon systems.

Indeed, it may be that we already have overclassified in this regard. It is not so harmful that someone in the Soviet Union, by reading a U.S. journal, might conjecture that we are working on a given weapon. What is more important is that we develop that weapon quickly and reach superiority through its availability to our military forces. It undoubtedly pays off for the Soviet Union to maintain a large effort devoted to obtaining information from the United States. But it does not follow that because a payoff exists for them, it is advantageous to us in either timing or cost to add more impediments in information flow in an endeavor to handicap them. Because the Soviet Union is producing more engineers and scientists than we are, it is paramount that we move faster and be qualitatively more successful. No matter how thorough their spying, anything the Soviet Union gets from us can only later translate itself into military application. By the time they are able to act on the knowledge they have gathered, we should be working on the next generation of weapons—which we can do if we concentrate on broadening our body of skilled scientists and engineers and on choosing the right projects.

It is of some pertinence to note that in civilian products the Soviet Union's record in exploiting information from the outside has been pitiful. Managers in Russian industry are not rewarded for innovating. Instead, they are punished for slipping on production schedules, a likely occurrence if unfamiliar techniques of production or product modifications are introduced. The weaknesses of Soviet infrastructure and organization are immense. Shortages of virtually all materials, small standard parts, instruments, and routine equipment items are widespread. Much technical know-how that is readily available worldwide has not been applied in Soviet industry for lack of the mundane equipment and materials necessary to apply it.

In the United States, if a laboratory scientist needs 2 feet of tungsten wire, two-thousandths of an inch in diameter, and doesn't find it on the shelf, it can be quickly ordered from a catalog. If in a hurry, the scientist can make a telephone call and chances are the wire will arrive by air the next day. The Russian scientist with the same need would have to start a long process. The request would perhaps have to be entered in the next five-year plan for coal and iron (to make the steel for the machines to mine the tungsten), and this would lead only much later to wire-drawing apparatus and a planned incremental increase in the supply of tungsten wire—after the bureaucracy had gone through its allocation of available resources among many competitive requests. Such heavy bureaucracy and profound shortcomings of the infrastructure make it difficult for the Russians to utilize external information and ideas when they get hold of them.

Suppose we really could reduce to zero the direct availability to the Soviet Union of U.S. information and equipment potentially helpful to their military. What then do we do about our allies in western Europe? An effective NATO operation requires, to say the least, a rather free movement of equipment and information among the multinational forces. If we export civilian computer equipment to German, Dutch, or French firms, how do we arrange that such equipment will not be studied by Soviet agents in those countries or even be reshipped to the U.S.S.R.? We would have to insist that our friends in western Europe set up strong access controls similar to ours.[8]

It may be fairly straightforward for Germany to employ severe restrictions to prevent spying on a military unit. But what about a U.S. commercial computer purchased by a German company for use in ordinary civilian operations? How can we expect to persuade that company to treat the rooms where the equipment is stationed as classified areas and use burdensome procedures to clear each entering individual? When the computer is ready for replacement, how do we keep it off the used market or get the company to close that market to customers from potential enemy nations?

Over the last ten years, the United States has dispensed weapons and follow-up military services to well over 100 different nations. The total cost has added up to more than $120 billion. It has included some of our most sophisticated weaponry and has gone to some unstable nations like Iran. If we intend to keep up this level of flow of our military technology, then to attempt simultaneously a severe restricting of general technology transfer from the United States may mean only hampering our own internal weapons development.

The Universities and Technology Transfer

Some backward steps that will hurt our status badly in military technology in the long run are being urged on universities by some in the government. The

---

[8] Actually, an organization called COCOM (for Coordinating Committee) has existed since just after the end of World War II to control and prevent the transfer of sensitive technology from the NATO nations and Japan to the Warsaw Pact nations. However, it is a voluntary organization, tiny, poorly supported, and ineffective, with no means to force compliance.

idea is to apply access rules quite similar to the controls used for classified military information to some aspects of the universities' normal research and education operations. Now none can dispute that the Soviet Union is readily able to acquire basic science and engineering information at American universities. If the government of the United States wants to keep Soviet scientists and engineers from profiting by appearances on this country's campuses (in the belief that what they deposit here in knowledge and ideas is not an adequate quid pro quo), then the government should simply stop allowing them to enter this country.

Some have proposed that all foreign nationals be excluded from those educational courses and research programs in our universities whose content is potentially exploitable by the Soviet Union. But this would essentially prevent American universities from accepting any foreign students. The universities can hardly be expected to create a system to break down and analyze each lecture before it is presented and then cause the professor to separate and eliminate that material potentially useful to the U.S.S.R. They also cannot monitor the doors to all laboratories and classrooms to prevent entrance by all except those who have been preexamined and cleared for such entry. They cannot conduct security investigations of all foreign attendees (not to mention American citizens), including those, say, from western Europe, Asia, and South America, to make sure they are not working for the Soviet Union clandestinely.

Of course, we could deny all foreigners access to our graduate schools as enrolled students. But if this were done, the enrollment in engineering and physics would drop by a third. We already badly lack enough skilled instructors, and we are using foreign graduate students to teach basic courses, such as computer science, to our undergraduates. Thus our engineering schools would immediately have to drastically cut the number of entering American freshmen.

It even has been proposed that our universities should not engage in basic research that might be of aid to the Soviet Union. Those projects, it is suggested, should be carried out instead in the laboratories of private corporations, or independent nonprofit organizations, or the government itself, in facilities that are open only to cleared personnel. But we do not know ahead of time which research is sensitive and where the research might lead. It is a characteristic of frontier areas of science and engineering, the kind just right for research by universities, that the fundamentals and the applications are intertwined and the latter usually are not foreseeable. For example, the program of the 1940s to probe the physics of the flow of electricity in semiconducting solid materials could not have been compartmentalized and deliberately kept separate from the development of the transistor, probably the single most important electronics invention. The idea for the transistor entered the minds of its conceivers because, while doing research, they came to understand the basic physical phenomena involved and this comprehension pointed to the invention.

Research that is truly basic, unlike more practical industrial and military projects, cannot be pursued well in secrecy because it addresses problems too difficult for solitary, confined workers. Open exchange of ideas and exploratory findings is fundamental to progress. Requests to control information flow in certain highly specialized fields (like the development of codes and code-breaking technology, i.e., cryptology) are reasonable. Blanket restriction on all research and dissemination of results is not reasonable. It may be better to have the Soviets busy snooping and copying, while always following several steps behind, than to have them concentrating their resources fully on substantive scientific research which may cause them to come up with real surprises difficult to counter.[9]

Educational courses and research programs in American universities should remain open to all who meet the competency requirements for admission. Then our graduate students, professors, and research workers in universities throughout our country, in complete and free discussion with each other, stimulated by students and professionals from all over the world, will move more rapidly than those elsewhere to push forward the frontiers. This is just what we have done so superbly in the past. This is why twice as many Nobel prizes have been awarded to Americans in the last several decades as the combined total bestowed on all the rest of the world's scientists.

Admittedly, the Soviet Union will be helped by public dissemination of early experimental data. However, they will be on the outside, while our highly interactive insiders will stay continually ahead. When and if fundamentals are turned up that suggest possible applications to the military, well-separated project efforts then can be started by the Department of Defense and classified to any extent they desire. Access to work on those projects can be banned except as permitted by adequate clearance procedures applied individual by individual. Moreover, all documents can be controlled, and all apparatus kept in closed-off areas isolated from the university.

---

[9] Published estimates put the figure of Soviet scientists and engineers engaged worldwide solely in gathering technical information from other countries at 5,000.

The Role of Free Enterprise in Military Technology

The U.S. budget deficit has made it evident that the military weapons race with the Soviet Union has become unacceptably costly. Though very large, American resources are nevertheless too small to provide readily both for national security at presently increasing expense and the level of government services our citizens expect. Equally important, the U.S. technological industry cannot simultaneously maintain the large stream of advanced civilian products needed to ensure adequate economic superiority in an era of tough international competition while also providing an abundance of weapons systems that can outdo anything a potential enemy might invent.

Since we can't do everything, we have to choose carefully. We cannot afford mediocre performance in the selection of where to place our chips and in the thoroughness and competence with which we implement technological developments once the choices are made. As we spend the planned trillions for defense over the next decade, we had better make sure the new weapons systems are the right ones, that we have enough competent people to perform the work scheduled, and that the right match exists between technological advance and overall security strategy. Otherwise, the bigger the spending figure, the less credible will it be that we are building a strong defense. Instead, we will be in the position of merely frantically throwing money at the problem.

The science and technology to back up military requirements must be dominated by government, with appropriate roles for the private sector under government contract, but without reliance on free enterprise to call out the nation's defense needs, since an obvious conflict of interest precludes that. We have a mature system in the United States for government-industry cooperation to create technological systems for the military. It ranges from early research explorations and analytical studies of alternatives to full-fledged, multibillion-dollar projects in which prototypes are designed, built, and tested and procurement of production hardware quantities follows. Regulations abound to control competition for contract awards, account in detail for expenditures, and prevent waste and fraud. Nevertheless, the system is replete with shortcomings. Many are curable. To do a better job of selection, organization, and administration requires clarity of policy plus expertise in management. Highest among the requirements is the specialized skill to direct high-technology activities. This is in short supply in the whole country, but particularly so, compared with the severity and complexity of need, in the government.

The massive U.S. expenditures on defense include many billions of dollars annually for R&D and the production of technologically sophisticated hardware. Over a million Americans are employed in these endeavors. With so large an effort, it would be naive to believe that all projects in military technology deserve the sponsorship they get. Some are bound to have originated from errors of judgment and misinformation. Others might have been sound projects when first started but have been put out of date by unanticipated developments. Some have resulted from expert industry sales representatives, applying their art overskillfully to less expert decision makers and buyers within the government. Substantial duplication exists, not all accounted for by interservice rivalry and overlap, although both of these are substantial phenomena. Some suppliers that fail to do their tasks well are nevertheless carried by the government because of the dependence that develops on those sources. The procurement process itself does not always select the most appropriate supplier through competitive bidding. The fact that companies and whole communities often become dependent on specific military production contracts has a particularly harmful effect on the nation's overall economic performance. Few senators and representatives can consider voting against a weapons system if its cancellation would cause significant unemployment in the geographical areas where their voters are to be found. This is only one example of conflicts that all members of Congress find built into the matter of military procurement. Almost any senator or representative who is willing to work at it can come close to ensuring the continuation of a project in his or her area if it is seen as important enough locally. The legislator only has to make deals with enough colleagues to exchange votes in support of the others' "must" projects.

No substantial defense project is begun without some consideration of alternative approaches to meeting the requirement. Each such approach quickly becomes identified with competitive industrial contractors whose sales teams, unofficially but effectively, often include their senators and representatives, who in turn find accepted ways to importune the DOD as they seek to channel contracts to their constituents. (Most, but not all, members of Congress also spend much of their available time in efforts to understand what is really best for the nation.)

We cannot readily change this military procurement process without changing the way our democracy works. Still, the earlier a sound plan for procurement of military technology to meet a specific requirement is created in the Pentagon, the more promptly and certainly will all others in the act adjust to the indicated directions. Industry salespeople, Congress, local politicians, lobbying groups for the several military services, and the rest, when up against competent administration leadership that presents a strong and justifiable case for the correct decision, will be inclined to stop fighting it. As they work for higher employment in their companies or communities, they will turn to lobbying for participation in the right projects, those that serve the overall national interest.

On the other hand, if there is no clear set of signals, if national security planning is missing, weak, confused, or late—which is a common occurrence—then it should not be surprising to find a maximum of inelegant grasping for government money. The result of such exercises in self-interest is higher expenditures and less security per dollar of expenditure.

The Military's Problem in Technology Management

Now that we have explored some of the difficulties in using science and technology fully to attain a secure nation, let us return to the important problem of the providing for adequately skilled people to plan strategy and manage the implementations.

The military's plans for systems of advanced technology cannot be formed soundly without an understanding of things both military and technological. However, career officers need to do more than sit down with experts in science and technology from industry and, on the assumption that between them the two groups will understand everything, arrive at a list of the technological frontiers to be advanced and weapons systems to be procured. The government experts, military and civilian (largely in the Department of Defense but also in other spots, such as the Energy Department's nuclear group), have to be in the leadership position. The input from industry's people is essential, but it must always be remembered they have an ax to grind in their desire to win contracts. Much interactive, committeelike, joint effort naturally will take place and is permanently necessary, but there has to be overall systems leadership. That belongs with the government. Its personnel must do the integrating and deciding.

The trouble is that the government's staff, both career military and career civilian, is not strong enough—given the difficulty of the assignment—in the management of science and technology. Admittedly, in high-level appointive positions (in the Defense Department, for instance, among the under secretaries, assistant secretaries, and their principal assistants) will be found manager-engineers of top caliber. These usually are individuals who have left high-paying jobs in private organizations to serve their nation. They typically do so with great distinction, but only for a few years. And a large fraction of their time is taken up with selling and defending their programs before Congress. Also, included in the military-officer corps and the civil service are experienced people with technical postgraduate degrees, a liberal sprinkling of whom are superior management professionals. However, the total number of such personnel in government is too small and—the attractiveness of government careers for outstanding individuals in science and technology having suffered badly over the last few decades in comparison with the private sector—diminishing rather than rising.

A person who is unusually talented in directing technological effort can choose from many routes for career advancement. Private positions usually offer more opportunity to participate directly in scientific and engineering developments, and they provide greater financial rewards and peer acceptance. Government service for such individuals, both civilian and military, involves more bureaucratic frustration, less direct potential for personal satisfaction, and a lower standard of living for their families. Thus there are few generals and admirals with Ph.Ds.

In directing the development of military weapons systems, the government's project leaders experience greater demands than executives in technological industry. At any rate, the latter really work under the former's direction, because the DOD's project heads have the higher responsibility, namely, overall project success. And yet we find that the average quality of experience and innate management ability is weaker in the government staffs than in the industrial leaders who must be managed by those staffs. Only a fraction of those directing programs in military technology can do so with adequate poise, competence, and confidence when the requirements are so great.

The situation is not aided by a cultural pattern within the DOD that retires military officers at the peak of their ability to contribute, and a retirement-pay pattern that almost forces early retirement. It is a nationally disgraceful handicap to attaining excellent project management that high officers leave the service while still in their forties, and even the extremely exceptional individuals who attain the highest command responsibilities are gone in their early fifties. It is surely possible to redo the whole pay and retirement structure and add several more years to the period of availability of key talent. We surely can increase the attractiveness of government positions for the scientist or engineer suited to managing projects in military science and technology. The remuneration can be increased, and career paths for technical personnel in the Defense Department can be made firmer and clearer. Successful leaders of small technological projects should know that they will be given more important areas to direct next if they perform well. The technological route for advancing within the military forces should have its own system of rewards and acknowledgements. Career officers who are engineering managers should not feel that they must go off to command a military fighting unit somewhere in order to see their careers move forward satisfactorily.

Finally, it should be the policy that, in view of the importance of science and technology to the overall military task, a substantial number of the highest policy officers in the military services regularly will be chosen from those whose principal accomplishments have been in directing projects in military science and technology. Such personnel promotions have indeed been made in the past, but they have been relatively infrequent.

As the shortage of competent technological executives within the military has been growing, the engineering and physical science departments of American universities have been gradually becoming weaker, as earlier discussed. The simultaneity of these shortcomings in both the DOD and the universities suggests the need for a coupled action to help solve both problems.

The Department of Defense has for years subsidized an ROTC activity in some universities and postgraduate education for selected military officers. It should do much more. For instance, it should budget for an unusual kind of support of doctoral education in American universities. Outstanding students in science and engineering, those who obtain their first degrees with distinction, should be encouraged to apply for a new kind of career fellowship contract with the military forces. In return for a commitment to pursue a doctorate intensively and, when it is obtained, to become a military officer committed to several years in the service, the military should provide financial backing. The first graduate year might constitute a trial, a period after which either side could elect to bow out. Further, the military should offer parallel funding to the universities for each graduate fellow selected, to help provide the infrastructure for top-grade graduate study, such as modern laboratories and supplements for professors' salaries.

In this way the government would create the underpinnings of a strong future staff in science and technology within the military. At the same time, the DOD would ensure quantitatively and qualitatively adequate technological university activities in the United States, a fundamental base for national security in the long run. The U.S. military needs some 25,000 members with technical degrees, of whom 5,000 preferably should possess doctorates. If

1,000 of these Ph.D.s were to leave the service annually as they move along in their individual careers, the DOD would need an inflow of 1,000 from the universities yearly. An average of forty additional Ph.D.s finishing at each of twenty-five leading American universities, a program costing around $50 million a year, would provide the flow.

This $50 million, which amounts to 0.02 percent of annual defense expenditures, would save billions annually some years from now as a result of superior strategy decisions and management of the military's programs. Meanwhile, the strengthening of the engineering and science schools would aid the nation's civilian programs and international competitiveness as well. Higher profits in the technological industry, a probable result of more successful competitive performance in international commercial markets, would cause the government to obtain added revenues from taxes on earnings in amounts well exceeding the $50-million yearly cost.

Top Scientists' Contributions

Other approaches to enhancing the selection process should be tried. We should be more innovative in bringing the most brilliant scientific minds in the nation to bear on the decisions as to which frontiers to probe. The Defense Department has its science advisory boards, and it places research and study contracts with the technological industry, the universities, and private nonprofit organizations. Thus it might appear the DOD already is calling fully upon the scientific and technological talent of the nation. However, there are important limitations to present efforts which can be overcome.

The nation's preeminent pure scientists are individuals who seek fundamental knowledge without concern as to its application and are mainly to be found in our universities. Few of them are today involved in defense work. Most prefer it that way because, having the free choice, they opt to spend their time seeking to make fundamental contributions to our store of scientific facts rather than working out solutions to practical problems.

Still, most of these same researchers will grant that national security for the United States is important to all its citizenry. They certainly have an awareness that the scientific breakthroughs of today might exert an important influence on the military weapons of tomorrow. While they cannot (and probably preferably should not) be recruited for direct, detailed involvement with military technology, they might be tapped effectively for occasional but vital advice on scientific trends that military leaders might recognize as critical in the future. It is not to be forgotten that it was Albert Einstein who wrote to President Roosevelt to warn of the nuclear bomb potential, to recommend a crash program, and to express concern that an enemy nation might make the breakthrough first, to our fatal detriment.

Thus suppose that periodically a group of top basic scientific researchers from the universities were invited by the Secretary of Defense to be guests at some appropriate hideaway for a week. Here they first would listen to military experts outline the problems of national security as they envisage them for the decade ahead. These presentations would call out the desirability of being able to accomplish certain tasks whose attainment appears to be beyond present scientific knowledge. Also, they might describe some anticipated innovative employment of new science by our potential enemies which might conceivably lead to our undoing; it would thus appear important for us to develop counters in these areas. Next, the invited scientists and the experts on military requirements would interact in relaxed discussion. Possibilities of scientific and technological advances would be cited by some, their proposals dissected and embellish by others. Ideas put forth informally would stimulate the beginnings of serious inventions or definitive evaluations.

The scientists would leave with a greater appreciation of the challenges the military field offers in the use of science and technology to provide for the nation's security in a dangerous world. They would have observed that in many instances they were unable to say confidently whether nature, as we understand its laws today, allows or precludes accomplishing what the military says would be advantageous. The curiosity of the scientists would be piqued by their being unable to decide some of these issues immediately, and this might spark some desire on their part to contemplate and define the research necessary to settle the scientific fundamentals involved, even though that research would not be theirs to perform. For their part, the military would obtain from such an exercise a much broader feel for the horizons of modern science. They would adjust their priorities and ponder possibilities that otherwise might not have entered minds.

The interchanges would elevate the level of communication between the military and the most outstanding research scientists of the nation, and this would have favorable indirect consequences as well. We have to assume that applying science to military problems is permanently a priority area of national endeavor, and that wide U.S. lead in such endeavors will come closer to guaranteeing a type of world U.S. citizens, including the pure scientists themselves, would prefer than if the Soviet Union were to become dominant. In the 1950s the science underlying military needs received more attention from the best university minds than it has in recent decades. Perhaps back then some preoccupation with military-related science occurred in academic circles. However, we since have moved steadily in the other direction and have gone too far.

Our military technology programs are imbalanced toward mere incremental improvements, and there is too little striving for the big steps that new scientific discoveries make possible. We are wasting resources and realizing less security through our failure to inject enough scientific creativity into military programs. This waste uses up available resources, and there are fewer funds left over nationally to support the very explorations into the fundamentals of pure science that talented university researchers would like to pursue.

The Triangle and National Security

We started this chapter with the thought that, by criteria most Americans would set, the first priority in the employment of science and technology should be to aid in preserving our freedom. If we do not utilize these intellectual disciplines effectively in the number 1 priority area, considerable doubt would appear justified about our ability to use science and technology well for any other facet of the national interest. If security is the first priority, then study of this application of science and technology should particularly show how society, technology, and liberty are interconnected. What we in the United States do with our options for the pursuit of security will have a great influence on the probabilities of world war or world peace.

The national security triangle of society-technology-liberty, when properly drawn for America, discloses at least three givens that we must accept, like it or not: (1) The society focus of the triangle will not in the foreseeable future produce actions to rule out all wars. (2) The liberty focus indicates that our freedom will remain in danger. (3) The technology focus will continue to add perils.

Our resources are finite. Nevertheless, we shall insist on being secure, so our nation must set security strategies. No matter how innovative those strategies, they must include the ever-continuing development of superior technological weapons. For such developments to occur effectively, strong and steady national program of science and engineering education and research are mandatory. While cooperation with allies must be maintained, this will prove increasingly more difficult, and we shall be forced to go it alone in some critical phases of defense. Funding for national security purposes thus will remain a major part of the government's budget, which will be even larger because much unprofitable dissipation of effort and resources is bound to continue. Since our "selfish constituency" approach to democracy seems permanently dear to our hearts, we are forced to countenance its equally permanent resultant waste even as we enjoy its contributions to our sense of liberty.

If we accept all these conditions as unavoidable, we must adopt the long-range view that national security will remain the priority issue for America. However, the more competently and imaginatively we perform in meeting this prime requirement, the better chance we have to keep it from being totally dominant. Then adequate resources will be available to provide the many other benefits science and technology can offer the society.

Index

About the Author

Simon Ramo is a cofounder and director of TRW Inc., one of the world's largest high-tech companies, and chairman of the board of the TRW-Fujitsu Company. A winner of the Presidential Medal of Freedom in 1983 and the National Medal of Science in 1979, he was TRW chief scientist for the nation's Intercontinental Ballistic Missile Program and was chairman of the President's Committee on Science and Technology under President Ford. More recently, he served as chairman of President Reagan's Transition Task Force on Science and Technology. He is a visiting professor at the California Institute of Technology and a fellow of the faculty of Harvard's Kennedy School of Government.

Dr. Ramo is the author or coauthor of many books, including *Peacetime Uses of Outer Space, The Management of Innovative Technological Corporations, Century of Mismatch, America's Technology Slip, Introduction to Microwaves, Fields and Waves in Communications Electronics,* and *Extraordinary Tennis for the Ordinary Player.*